METAL

METAL
DESIGN AND FABRICATION

David and Susan Frisch

PHOTOGRAPHY BY JOSHUA WHITE

WHITNEY LIBRARY OF DESIGN

an imprint of Watson-Guptill Publications/New York

Senior Editors: Roberto de Alba, Ziva Freiman
Editors: Micaela Porta, Victoria Craven
Designer: Areta Buk
Graphic Production: Hector Campbell
Set in 11pt. Weiss

First published in 1998 in the United States by
Whitney Library of Design
an imprint of Watson-Guptill Publications
a division of BPI Communications, Inc.
1515 Broadway, New York, New York 10036

Library of Congress Cataloging-in-Publication Data

Frisch, David.
   Metal : design and fabrication / David and Susan Frisch ;
photography by Joshua White.
      p.   cm.
   Includes bibliographical references and index.
   ISBN 0-8230-3034-2
   1. Metal-work. I. Frisch, Susan. II. Title.
TS205.F74 1998
671—dc21   98-7883
                               CIP

Printed in Singapore

First printing, 1998

1 2 3 4 5 6 7 8 9 / 06 05 04 03 02 01 00 99 98

ACKNOWLEDGMENTS

We would like to thank the following people and organizations for their assistance in compiling this book:

Artistic Sandblasting, Los Angeles, CA
A + M Welding, Gardena, CA
Accurate Alloys Inc., Irwindale, CA
Ace Metal Spinning, Los Angeles, CA
Azusa Tube Bending, Azusa, CA
Bronze-Way Plating, Los Angeles, CA
Areta Buk
Burbank Sheet Metal, Burbank, CA
Charles Caine Co., Los Angeles, CA
Victoria Craven
Roberto de Alba
Ziva Freiman
Industrial Metal Supply, Burbank, CA
Dan and Ellen Kolsrud
Michael P. Krival, Esq.
Franz E. Kurath
Lane + Roderick, Santa Fe Springs, CA
Mechanical Metal Finishing, Gardena, CA
Octavio Mora
Pima County Air Museum, Tucson, AZ
Micaela Porta
Tube Service, Santa Fe Springs, CA
Western Glass, Vernon, CA

CONTENTS

PREFACE 9

CHAPTER 1	**PROPERTIES OF VARIOUS METALS**	**11**
	ORIGINS	11
	IRON AND STEEL	11
	ALUMINUM	16
	COPPER, BRASS, AND BRONZE	17
	LEAD	19
	TITANIUM	19
	GALVANIC CORROSION	19

CHAPTER 2	**STANDARD METAL PRODUCTS**	**21**
	SHEET PRODUCTS	21
	EXTRUSIONS	24

CHAPTER 3	**CUTTING AND SHEARING**	**31**
	SHAPE CUTTING IN SHEET AND PLATE	31
	CIRCLE CUTTING	38
	CUTS IN BAR STOCK AND OTHER EXTRUSIONS	39

CHAPTER 4	**DRILLING AND PUNCHING**	**42**
	DRILLING	42
	PUNCHING	48

CHAPTER 5	**BENDING AND ROLLING**	**50**
	BENDING EXTRUSIONS	51
	BENDING SHEET PRODUCTS	54
	ROLLING SHEET AND EXTRUSION	55

CHAPTER 6	**PRESS WORK AND SPINNING**	**58**
	PRESS WORK	58
	SPINNING	63

CHAPTER 7	**MILL AND LATHE TECHNIQUES**	**67**
	MILL AND LATHE TECHNIQUES	67
	THE MILL	67
	CHEMICAL MILLING AND ETCHING	73
	THE LATHE	73

CHAPTER 8	**WELDING AND SOLDERING**	**78**
	WELDING	78
	SOLDERING	86

CHAPTER 9	**MECHANICAL FASTENING**	**87**
	THREADS	89
	SET SCREWS	92

BOLTS		92
MACHINE AND CAP SCREWS		92
SECURITY SCREWS		93
NUTS		93
WASHERS		94
STUD WELDING		95
RIVETS		96
PINS		97

CHAPTER 10	**BLENDING AND GRINDING**	**98**
	PATTERNED TOOLS	98
	ABRASIVES	101

CHAPTER 11	**SANDBLASTING AND DEBURRING**	**106**
	BLASTING	106
	DEBURRING	108

CHAPTER 12	**BRUSHING AND POLISHING**	**111**
	BRUSHING	111
	POLISHING	113
	SPECIFICATIONS	115

CHAPTER 13	**ORGANIC COATINGS**	**117**
	SURFACE PREPARATION	118
	PRIMER	119
	CLEAR FINISHES	119
	APPLYING ORGANIC COATINGS	120
	VITREOUS ENAMEL	123

CHAPTER 14	**PLATING AND ANODIZING**	**124**
	PLATING	124
	ANODIZING	129

CHAPTER 15	**STRATEGIES FOR CREATING VOLUMETRIC SOLIDS**	**131**
	FRAME-SUPPORTED SHAPES	131
	BUILT-UP SHAPES	136

CHAPTER 16	**DETAIL AND VISUAL SCALE**	**138**
	WORK WITH EXISTING PRODUCTS	138
	EMPLOY PRECISION ONLY WHEN NECESSARY	139
	LARGE-SCALE ENGINEERING SOLUTIONS ARE OFTEN SCALABLE	141
	CONSIDER HOW THINGS THAT ARE TOUCHED DIFFER FROM THINGS THAT ARE VIEWED	142
	FOCUS YOUR INTENTIONS	143

CHAPTER 17	**MOTION**	**144**

CHAPTER 18	**MOVEMENT OF HEAVY PIECES**	**153**

CHAPTER 19	**COMPUTER AND MACHINERY INTERFACES**	**158**
	COMPUTER NUMERICAL CONTROL	158
	NESTING AND BENDING	160
	DESIGN IMPLICATIONS	161

GLOSSARY OF TERMS		163
INDEX		175

PREFACE

We chose to write this book to try and bring together two worlds we straddle in our own lives. We are both designers with degrees in architecture, and we are also manufacturers, operating a metal fabrication shop where we build metal furniture and architectural components for our own projects as well as for other designers. Our purpose here is to acquaint designers, architects, and sculptors with the basic processes available to them in metal fabrication, so that they can begin their design process with a sense of how their work might be built. The book is not intended to instruct in the use of machinery, but rather to provide a framework of possibilities for people who design custom metalwork. By "fabrication," we are referring specifically to the assembly of items from stock metal components (sheet, plate, and extrusions). We have favored those processes that are readily available and economical for custom work or small runs of parts rather than processes used for larger industrial endeavors, such as auto assembly.

Two areas of metalworking are excluded from the scope of this book, as each is worthy of an entire book of its own. The first is casting, which is the process of pouring molten metal into hollow molds, where it solidifies into the desired shape. Casting is very well suited to the making of complex shapes, many of which cannot be produced from stock components. The second is the set of traditional processes commonly referred to as "blacksmithing." Blacksmithing is alive and well, practiced by craftsmen employing methods refined by millennia of experimentation and inspiration. However, metalworking is a craft that abruptly altered course with the harnessing of electricity. The new processes developed in the last century, coupled with the common availability of stock metal components, have profoundly changed the way metal parts are assembled and detailed. Nevertheless, for all that has been gained from the new processes, other things have been lost. The elegance of detailing present in traditional metalworking is often missing in the translation from older to newer techniques.

Craft exists in an uneasy partnership at this time with art and architecture. In the arts, enormous value is placed on personal expression; most designers and artists abhor the idea of relying on traditional motifs in their work, unless done so ironically. The perpetual revolution of the arts is almost the exact opposite of the continuum of craft. Experimentation is certainly a part of craft, but so is the learning and acceptance of certain pragmatic facts about the making of things. These facts—*how* things are actually made—must be a part of the process of designing *what* will be made. Without some mutual understanding between designer and fabricator, it is unlikely that anything profound can really result from their efforts. It is our hope that this book will serve as a point of departure for the initiation of that dialogue.

Susan + David Frisch
Burbank, California
1998

CHAPTER 1
PROPERTIES OF VARIOUS METALS

ORIGINS

The earliest metals used by humans for tools and decorative implements were gold, silver, and copper. Prior to millennia of mining and scavenging, these metals were probably found in some abundance in their pure crystal form on the surface of the earth. Gold and silver are too soft to make very useful tools, but pure copper can be hardened by hammering into a serviceable cutting tool. Copper also has a low melting point, which can easily be reached with a wood-fueled fire.

Prehistoric metalworkers made an extremely critical discovery when they found that certain ordinary looking rocks found near copper deposits could be crushed and scattered into a hot fire, and that traces of copper in the rock would coalesce into a puddle of molten metal. This method of obtaining metal is called smelting; rock that contains a mixture of the desired metal and various impurities is called ore. Further experimentation led early metal smelters to add traces of zinc, tin, lead, and other admixes to the copper, resulting in new combinations of many metals, called alloys. An alloy is a combination of a base metal with traces of other materials, which substantially alter the physical properties of the pure base. Some of the first alloys were bronze and brass, both made primarily of copper.

IRON AND STEEL

Iron, which superseded bronze as the metal of choice for tools, does not exist in abundance in its pure state in the earth's crust. The largest source of pure iron is meteorites, which could explain why some early names for iron translate roughly to "star metal" or "stone from heaven." Meteor deposits are hardly a reliable source for iron, however, and at some point, humans began smelting iron from ore just as they had earlier smelted copper. Iron ore is extremely common, and massive deposits are found in many sites all over the world.

Iron can be smelted, slowly, from crushed ore in coal-filled fire pits. The process is substantially improved by feeding the smelting fire with a stream of air. This is called stoking the fire. Stoking was probably first accomplished by locating fire pits in naturally windy locations, or by fanning air into a hole at the base of the pit. Stoking was gradually improved with a series of inventions, such as blowpipes, air bladders made from animal skins, and bellows, all the way to the electrically powered, preheated air turbines used today.

In the Middle Ages, most iron was refined from ore using a crude version of a device now known as a blast furnace. Crushed iron ore and charcoal were deposited in a stone vault, the mixture was ignited, and fed by stoking with a bellows. The molten iron (called pig iron) would travel to the bottom of the furnace, where it was released through a tap into molds. Charcoal adds traces of carbon to the iron, which dramatically alters its workability. Crushed limestone was also added to the mix because it bonds with impurities in the metal to form a glassy slag that can be skimmed away. Until the late nineteenth century when the use of coke replaced charcoal, the use of wood to produce charcoal for iron production was one of the main causes of worldwide deforestation.

Some of the iron produced in this way was almost pure and very malleable, primarily because it did not contain much carbon. This material was called wrought iron. Wrought iron was extremely useful for a wide variety of tools and weapons, and for centuries it was the most commonly used metal for these items. Much of the iron produced in blast furnaces, however, contained large amounts of carbon (up to 4 percent). This material, called cast iron, is very brittle and cannot be worked once it solidifies. Occasionally, a small amount of a malleable iron alloy was produced in the furnace. This could be made hard by heating and then rapidly cooling in water or oil. The material, which was close to what we now call steel, contained about 1 percent carbon.

Until the middle of the nineteenth century, steel was reserved for very specific applications, such as the cutting edges of swords, shears, and clock springs. Wrought iron was used for almost everything else. The quality of early steel was fairly unpredictable, and methods developed for creating quantities of homogeneous steel were closely guarded secrets. Some of the best steel in the world was made in very small batches in steelmaking centers such as Damascus and in parts of Asia, where metalworkers developed highly sophisticated processes for making steels with a variety of unique and desirable qualities. Some of these processes were never truly duplicated in the West.

To make wrought iron in large quantities, processes were developed in Europe after approximately 1300 A.D. that involved the remelting of a mixture of pig iron (made in blast furnaces), iron scrap, and charcoal (or later coke) in a hearth or furnace. The air forced into the melt to increase the temperature of the fire would also cause the oxidation of much of the carbon in the mix, along with other impurities such as silicon. Steel was first produced in large quantities in much the same way, but either the carbon was prevented from completely oxidizing in the pig iron, or it was actually added to molten wrought iron.

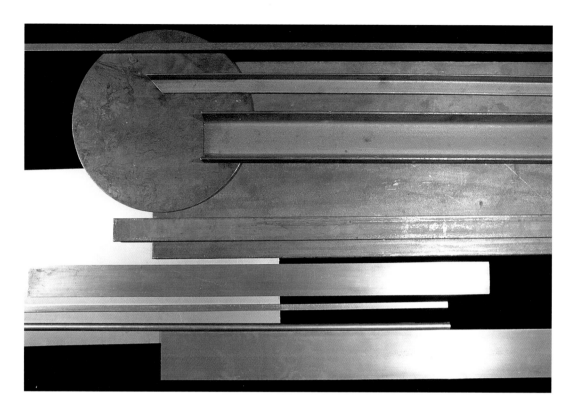

Hot-rolled (top) and cold-rolled carbon steel.

Modern steel is primarily made using one of two processes: the Basic Oxygen Furnace Process, and the Electric Arc Furnace Process. The Basic Oxygen method uses an oxygen lance to blow pressurized pure oxygen into a furnace containing scrap metal and pig iron. The Electric Arc Furnace Process uses a set of carbon electrodes to create an electric arc that is sufficiently large to melt the metal. In both methods, the composition of the steel is changed by removing unwanted elements from the mix while also adding desirable elements. Carbon is the most common additive. Steel also contains traces of other materials that are by-products of refining, such as sulfur and silicon. Materials added to alter the characteristics of different steel alloys include chromium, nickel, and manganese.

Most of the variations of steel produced today are referred to as Standard Steels. Standard Steels are categorized into carbon and alloy steels. Carbon steels are the most commonly produced kinds of steel and are used for most structural and mechanical applications. There are hundreds of specialized grades designed for very specific applications, and for large or critical jobs, you would certainly want to investigate which to specify as the best match to the final use. For the majority of fabrication work, however, the grades used are those designated with the numbers 1008 through 1020.

Alloy steels are steels with more than 1 percent carbon content, as well as steels that contain special additives. The most common alloy steels are called stainless steels. Stainless steels have superior resistance to corrosion, and are often stronger and harder than carbon steels. The corrosion resistance of stainless steels comes from the existence of a microscopic oxide layer that forms on

Stainless steel with various finishes.

stainless steel when it is exposed to the atmosphere and deters further oxidation. Chromium, a very hard metal, is added to most stainless steel alloys. Chromium content varies in different alloys from 4 percent to about 30 percent. The most common stainless alloy used in fabrication is alloy 304; it contains roughly 20 percent chromium, 10 percent nickel, and traces of several other metals. It is very weldable, and has good corrosion resistance. Alloy 303 is often used for parts that must be extensively cut or machined, since the addition of sulfur makes it softer. By and large, however, stainless steel is hard on cutting and grinding equipment, and is more expensive, pound for pound, than carbon steel. For these reasons, parts made from stainless are usually more expensive than identical parts made from carbon steel.

Another commonly used alloy steel is called high-strength, low-alloy steel, or HSLA.

The HSLA steels are also sometimes referred to as COR-TEN steel. COR-TEN steels have increased strength and the unique feature of enhanced corrosion resistance over carbon steels. When exposed to the atmosphere, COR-TEN steels will form a consistent oxide layer in a matter of weeks, preventing further corrosion. The oxide layer, which is usually a deep reddish orange, is very attractive and makes the product desirable for some architectural and sculptural applications.

Although steel is typically used "as is" directly from the mill (see Chapter 2), there are a number of ways to increase (or, occasionally, decrease) its hardness and strength. Some steels can be hardened by hammering or compression. This is called cold hardening. For most applications, however, steel is heated to achieve the desired change in its strength. At very specific temperatures, the crystalline structure changes dramatically, altering its strength characteristics. It is interesting to note that in the absence of thermometers, metalworkers throughout history have relied on the color of the glowing steel to indicate temperature. The process of heating steel to an almost liquid state, and then cooling it quickly in an immersion of water, oil, or air, is called quenching. Quenching creates a thin layer of metal crystals that are harder and more brittle than the rest of the part. Quenching can be followed by tempering, which is a reheating of the part (below the melting temperature) to produce crystalline structures in the part with additional strength or ductility. Annealing is the process of heating the steel and then cooling it very slowly. The slowness of the cooling creates a softer, uniform, crystalline structure; annealing is done to make hard steel softer for cutting and forming. Once the desired operations are performed, the annealed piece can be hardened again by quenching. As a group, these processes are called heat treating. Most fabricators do not heat treat parts themselves; parts are sent to a vendor specializing in the process.

Another process used to harden the surface of a steel part is called case hardening. Case hardening is accomplished either by changing the chemical composition of the surface prior to heat treating or by heating only the surface of the part with a torch or an electric coil placed around the part. Case-hardened parts are often sprayed with water or oil to cool the surface at the desired rate.

One other major category of steel, not included in the Standard Steels, are a set of alloys called tool steels. Tool steels are specifically intended for parts like hammer heads and plows that will be subjected to extreme stress. Tool steels are categorized as oil, water, or air hardening, depending upon the quenching process best suited to the particular alloy.

Steel has gradually replaced wrought iron as the final product of iron smelting worldwide. Actual wrought iron is very difficult to find, sold only in small batches through specialized sources. When the trend in industry began to favor steel around 1910, it was not without substantial displeasure on the part of the craftsmen called blacksmiths, who made most of the utilitarian metal implements required by society. The workability of carbon steel was inferior to wrought iron when using traditional fabrication processes that rely on hot working (changing the shape of pieces by hammering and beating after they have been heated until soft). Over thousands of years, blacksmithing developed a vocabulary of tools, shapes, and methods of assembly based upon traditional energy sources such as water power and charcoal. The invention of the electric motor and the switch to carbon steel as the the primary metal for fabrication have profoundly changed that legacy. The new palette of tools, shapes, and methods of assembly acquired in this century is the subject of this book.

Aluminum is another metal that is commonly used for fabrication. Aluminum has a light gray color, lighter than that of steel or stainless steel, and is not found in its pure state on the earth's surface, but only in the form of compounds. It was not commercially viable to smelt aluminum until the nineteenth century, when a material called bauxite was discovered in France. Bauxite is an aluminum ore that is found most frequently on the earth's surface in tropical climates. It contains a substantial amount of aluminum, trapped in the form of aluminum oxides, such as alumina.

Pure aluminum is collected by separating the alumina from the bauxite by heating the crushed ore in liquid and skimming the dissolved alumina from the surface as impurities sink to the bottom. The alumina is then placed in a device called a precipitator, which causes the dissolved alumina to form into tiny crystals. The crystals

Aluminum, various finishes, extrusions, and plate.

are removed from the liquid and dried in kilns to produce a pure alumina powder. The powder is then placed in a series of tanks called reduction cells. In the reduction cells, the alumina is mixed with another compound in a liquid state and charged by an electric anode. The oxygen in the alumina bonds with the anode, and the pure aluminum collects as a crust on a cathode on the bottom of the cell. When you compare this method with the more intuitive process used for thousands of years to harness iron, it is easy to understand why aluminum was not used until very recently.

Aluminum is non-magnetic, non-sparking, and about one-third the density of steel. In its raw state it is very soft (like copper) and can be welded, although greater technical skill is required than with welding steel. Aluminum components are frequently hardened by heat treating, just as steel can be. Aluminum is very reactive with oxygen: Aluminum parts form a microscopic layer of oxide on their surface when they are exposed to the atmosphere. This layer makes aluminum very resistant to weathering, but it can be corroded by alkalies. It is also extremely conductive, and this is why it is often used for carrying electrical current. Highly reflective, it can reflect up to 90 percent of radiated heat. For this reason, it is often used for insulation, radiators, and packaging.

Aluminum is purchased in a wide variety of alloys that are designed for specific functional purposes. The alloys are assigned numbers in a series of seven categories, from 1000 to 7000. This number is followed by a letter code that indicates the tempering characteristic of the particular alloy batch. The most common tempering category found in aluminum products is "T" (typically T-1 through T-10), which means that the aluminum has been heat treated to produce a stable level of tempering. Far and away the most common aluminum alloy is 6061-T6, which is a member of the 6000 series. Because 6061 contains traces of silicon and magnesium, it is suitable for heat treating. Also easy to weld, 6061 is commonly used both for fabrication and aluminum part casting. Two other alloys, 2024 and 7075, are often used in aviation because of their great strength.

COPPER, BRASS, AND BRONZE

Both brass and bronze are alloys of copper. The difference between brass and bronze is not precise; copper/zinc alloys that produce bright yellow metals are usually called brass, and other alloys, especially those that produce more reddish or brown metals, are usually called bronze. The metals added to copper to create alloys include zinc (usually for brass), tin, aluminum, silver, and nickel. Different alloys have differing physical properties that are suitable to particular tasks.

The main functional advantage of copper alloys over ferric (iron-based) metals is their resistance to corrosion. This makes them suitable for any application where contact with water is frequent and severe;

Copper and brass.

plumbing and marine hardware are two good examples. Copper is also extremely soft and easily hammered or formed into shapes. Brass and bronze are harder, but still far softer to work than steel.

The most common brass alloy is yellow brass (70 percent copper, 30 percent zinc). Yellow brass is made into most extrusion shapes (see Chapter 2). Yellow brass is also commonly made into two very unusual stock profiles, half oval and half round. Another alloy is red brass (85 percent copper, 15 percent zinc), which is only made into pipe. Brass comes in a variety of tempers, categorized as soft, half-hard, and hard.

Bronze also comes in a wide variety of alloys. Aluminum silicon bronze is used for high-strength applications. Aluminum-nickel bronze is used for very high-strength applications, like shafts. Silicon bronze is almost as strong as mild steel, and is used primarily for marine hardware. It is also one of the easiest copper alloys to weld. There are other bronzes specifically suited for boat propellers, bearings, pressure vessels, high-temperature conditions, and wire. If your primary criteria is aesthetic, it is advisable to get samples of various bronzes to differentiate between their different base colors, which vary from yellow to orange to tan.

Another important factor in using copper alloys is their ability to take oxidized patinas that create a wide variety of beautiful surface colors. An excellent resource on this subject is a book called *The Colouring, Bronzing, and Patinization of Metals* (Hughes and Rowe, 1982, Watson-Guptill Publications). The book is something of a "cookbook" for designers and fabricators who wish to achieve specific patinas.

LEAD

Lead is a soft, dull-gray metal that is very soft and dense. It is used to weigh things down, to shield against radiation, and to provide soundproofing in walls and ceilings. Lead is too soft for many of the fabrication processes described in this book, and its toxicity makes it unsuitable for many purposes that it might otherwise serve. Lead is produced in rolled sheet, ingots, and shot. It should not be handled without gloves.

TITANIUM

Titanium was discovered as an oxide in 1791, but was not obtained in its pure form until 1910. It was not commercially viable to produce it until 1946, when a refinement process was discovered. Titanium is roughly equal in strength to steel, but only half as heavy. It is used in applications requiring both high strength and lightness, such as in parts for aircraft. It is extremely resistant to many things that can destroy other metals, such as oxidizing acids, seawater, and chlorine gas. Because of its corrosion resistance, it is frequently used for surgical components and tools. Titanium is also nonmagnetic.

Titanium, however, is about as difficult to work as stainless steel. It will harden at the point of contact with a cutting tool, so it should be cut at high speeds with very sharp implements. It is very easy to weld, but the weld zone must be completely shielded from atmospheric impurities. Although titanium is extremely expensive, its unique combination of strength and lightness makes it indispensable for many applications.

GALVANIC CORROSION

It is important to keep in mind that dissimilar metals, when in contact with each other in an electrolytic solution (like moisture), will develop negative potentials that can generate a weak electrical current. This flow of current slowly corrodes one of the metals. Metals are categorized by their likelihood to corrode when exposed to this condition (see Galvanic Chart on page 20). The higher a metal is on the list, the more likely it is to corrode. Also, the farther apart the metals are from each other on the list, the more likely they are to react with each other, with the lower metal losing the battle. So, for example, on a sculpture made of aluminum and

Oxidation (rust and mill scale) on hot- and cold-rolled carbon steel.

copper (in direct contact), the aluminum will slowly corrode unless it is separated from the copper. Dissimilar metals are often separated using non-metallic barriers, such as neoprene, felt, and plastic washers and spacers.

GALVANIC CHART		
	Most likely to corrode:	magnesium
		aluminum
		zinc
		cast iron
		carbon steel
		304 stainless
		tin
		lead
		nickel
		brass
		bronze
		copper
		titanium
		silver
		graphite
	Least likely to corrode:	gold
		platinum

CHAPTER 2

STANDARD METAL PRODUCTS

It is breathtaking to consider that until the mid-nineteenth century, most metals were only available to metalworkers in their raw state, formed into crude ingots. If a metalworker needed, for example, $1/4$"-thick steel plate, he had to heat and hammer a bar to the desired thickness. Large plates had to be made by joining many smaller ones together. Tubing and pipe was made by laboriously wrapping flattened bar around a solid core and welding it together, inch by inch.

Today, stock metal products are manufactured in a wide array of shapes and sizes, ready for assembly. The two basic categories of metal products are sheets and extrusions. Sheets are flat panels, graded by their thickness. Extrusions are lengths of metal, formed to take a particular profile. The range of product choices is slightly different for each metal, so it is important to gain some familiarity with what is generally available and what is not.

SHEET PRODUCTS

Sheet products are formed in one of two ways: They are hot rolled or cold rolled. Hot-rolled sheet is formed by squeezing hot metal through a series of increasingly narrow pairs of smooth rollers that thin the raw stock to a desired thickness. Hot rolling is primarily used for carbon steel and stainless steel, since their hardness makes it easier to work them at high temperatures. On carbon steel, the hot rolling process leaves a rough layer of bluish oxide on the surface of the sheet. This layer can be removed chemically (by a process called pickling), or by sandblasting.

Cold-rolled sheet is also made by squeezing the metal through rollers, but in this case the metal is at room temperature. Tremendous pressure is required to deform the unheated material. The advantage of cold rolling is that it produces a very smooth, dimensionally accurate product, free of oxide buildup. Better quality steel and stainless steel is produced this way, and all softer metal sheet is made using cold rolling. Cold-rolled stainless sheet is commonly available in a finish called #2b, which arrives with a smooth, bright finish

Sheet and plate.

and is excellent for welding and decorative treatments. Another common finish on stainless sheet is called #4, which is a polished finish overlaid by very evenly grained linear brushing. This finish is commonly used on kitchen and bathroom equipment. Stainless can also be obtained in fully polished (mirror polished) sheets. Softer metals, like copper, brass, bronze, and aluminum, are available in smooth or even polished sheet as a standard product.

Sheet sizes available from end-sellers vary from metal to metal. Steel is usually available in seven sizes: 3' × 8', 3' × 10', 4' × 8', 4' × 10' (the most typical), 4' × 12', 5' × 10', and 5' × 12' sheets. Copper, brass, and bronze are available erratically in 2' × 8', 2' × 10', 3' × 8', 3' × 10', 4' × 8', and 4' × 10' sheets, depending on thickness. Aluminum is stocked in 4' × 8', 4'x 10', 4' × 12', 5' × 10', and 5' × 12' sheets. Thin sheet in all metals is also available directly from the mill in coils. Sheet size is an important limitation to consider in the design of metal elements; joints must be placed in anticipation of maximum sheet sizes. It is sometimes possible to order metal in larger dimensions than normally stocked, but this order must be placed directly with a mill, and usually a very large minimum order is required.

Sheet products are graded according to thickness, and every metal has a different system used to designate thickness. With steels, thickness is calibrated to a series of gauges, running (hypothetically) from 52 to 1 gauge. In actual practice, most steel suppliers stock thicknesses between 28 and 7 gauge. The lower the number, the thicker the material. Thus, 28-gauge steel is approximately .0149" thick, while 7-gauge steel is approximately .1793" thick. Sheet is usually intended to function as a "skin" in fabricated objects. Flat steel that is intended for structural applications is called plate, and plate is described by its actual thickness, such as $1/8"$, $3/16"$, $1/4"$, $3/8"$, $1/2"$, $3/4"$, and so on. The two systems overlap slightly at 11-gauge (just under $1/8"$ thick).

Most copper sheet is broken out by its weight per square foot rather than its thickness. A 16-ounce copper sheet is .0216" thick, thin enough to be easily bent by hand. A 185-ounce sheet is $1/4"$ thick. Above this thickness, copper plate is listed by its actual thickness in fractional form ($1/4"$, $3/8"$, $1/2"$, up to 3"$).

Brass and bronze are stocked by gauge, by fractional thickness, and even by decimal thickness, depending on the alloy.

Aluminum sheet and plate are designated by their actual thickness, but the dimensions are called out as decimals (.032, .050, .190, etc.). The availability varies depending on the particular alloy that is selected.

Aside from smooth plate, some mills produce a product called diamond plate, which is embossed with a series of crisscrossing diamond-shaped welts. Diamond plate is intended for use on surfaces on which people walk, where a smooth finish could be slippery and dangerous; examples include steps, ramps, and platforms. Diamond plate is available in steel, stainless steel, and aluminum, usually in $1/8"$ or $1/4"$ thicknesses.

There are a wide variety of "after market" products made from sheet for different purposes. One of the most common of these is perforated metal plate, called "perf" for short. Perf sheets are available in numerous hole sizes and shapes (round being the most common). The holes are arranged in staggered or straight patterns punched

A stack of expanded metal sheets.

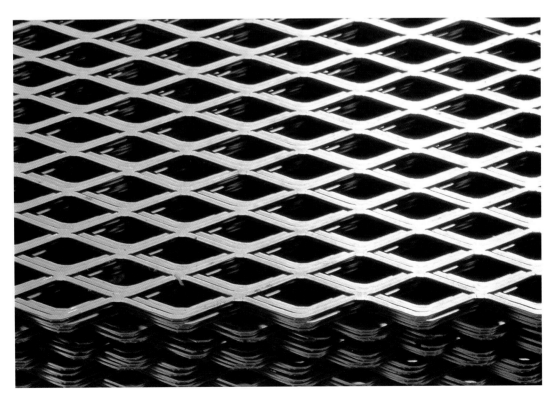

Perforated sheet with expanded metal in the background. The circular sheet in the foreground has been rolled (see Chapter 5).

out of stock sheet by special shears. Another product, called expanded metal, is made by slitting and then stretching sheets to create a three-dimensional lattice work. Other products having custom patterns embossed in their surfaces are too numerous to catalog accurately.

EXTRUSIONS

Extrusions are metal bars with a definitive profile in cross section created by drawing the metal through dies or rollers that are shaped like the profile. Extrusions include structural and decorative profiles, solid bar stock, pipe, and tube. Different metals are available in different ranges of extrusions, depending on their intended use. Like sheet, extrusions are formed by both hot and cold rolling.

STRUCTURAL EXTRUSIONS

Structural extrusions include channels, tees, I beams, and rail profiles (used to make train track). "Zees" are sometimes listed in textbooks, but in practice they are infrequently encountered. Structural shapes are made primarily from steel and stainless steel and have rough surfaces resulting from hot rolling. Structural shapes have tapering cross sections that tend to thicken at joints. For example, on channels, the two parallel edges (called flanges) thicken as they approach the side perpendicular to them (called the web). Structural lengths meant for use as beams are also given a slight bend along their length, curving in the direction that they are expected to be loaded. In this way, the extrusion will bend down into a level position instead of sagging. This property is called camber. Camber can most frequently be observed on flatbed trucks, which are usually designed to bow upwards when they are empty.

"I beam" is a generic term used to describe a number of structural sections that have a profile shaped like a serif capital letter "I." The vertical line is the web, and the pair of horizontal lines are the flanges. The actual categories used for these sections are W, M, HP, and S shapes.

Tee sections are, as their name implies, tee-shaped. Larger tees are made by splitting I beams. Angles are shaped like a capital letter "L." The sides of an angle are called legs, and the point of connection is called the heel. Angles are available in many sizes with equal legs, and in some sizes with unequal legs. Channels are C-shaped extrusions with a matched pair of legs pointing in one direction from the flange. A zee is much like a channel, but the legs point in opposite directions.

Most structural extrusions are stocked in 20-foot lengths, although I beams are often ordered for specific jobs in longer lengths. While the majority of structural shapes are made from steel, there is also a limited selection of structurals available in aluminum. With aluminum,

Standard solid extrusion profiles. Top row, left to right: C-channel, I-beam or column, tee section, angle. Bottom row: round bar, square bar, hexagonal bar, flat bar. Half-rounds and zee bars are not frequently encountered.

there is also another category of shapes available called architectural
extrusions. Architectural extrusions take the same configuration
as structural profiles, but they possess sharp (90-degree) corner
intersections, smooth surfaces, and joints, legs, flanges, and webs
of even thickness. They are used for applications such as window
frames, trim, and furnishings.

Thin steel sheet is also formed to create shapes used in construction.
Although these are not technically extrusions, they are sold in
lengths. Examples include metal studs, hat channels, and door and
window mullions.

SOLID SHAPES

Solid bar shapes include round, square, rectangular (called flat bar), and
hexagonal. Half rounds and triangular bars have been manufactured
in the past, but are infrequently stocked today. All metals are available
in bar stock, although the selection varies. Virtually all metals are
available in wire, also considered an extrusion. Carbon steel, stainless
steel, and aluminum come in the widest variety of solid bar profiles.

Various solid bars with section cuts.

Solid bar in the rack.

In carbon steel, solid extrusions are available in hot-rolled and cold-rolled forms. Hot-rolled bar is usually rough, and has rounded edges and imperfect dimensional qualities. Cold-rolled bar is smooth and cleanly surfaced, and has square corners and a very accurate profile. It tends to be a little harder than hot-rolled bar, although the outer layer of oxide on hot-rolled is harder to grind. Stainless steel solid bar is most readily available in hot-rolled, or occasionally in sheared bar, which is made by cutting larger sheets into strips. Cold-rolled stainless does exist, but is expensive and difficult to find.

Copper is made into rounds, squares, rectangles, and a profile called bus bar, which is rectangular with rounded corners. Bus bar is used for making heavy-duty electric conductors. Bronze is made into a fairly limited selection of round, square, and rectangular shapes. Brass is made into a wider range of sizes and shapes than bronze, including hexagonal. Solid profiles are also available in unusual or custom profiles from manufacturers, most commonly in aluminum. Examples include window and door frames, and specialized trim and louver profiles.

PIPE AND TUBING

Pipe is typically round (occasionally square or rectangular) and is graded by a fairly complex system that begins with the approximate (nominal) inside dimension of the pipe (i.d.) for sizes up to 12" in diameter. Pipe of many sizes is then further classified into as many as three groupings based on the severity of its intended use: standard weight, extra strong, and double extra strong. Standard has the thinnest wall, double extra strong the thickest.

Another classification system, called schedule, is used for some pipe sizes to describe wall thickness. In pipe under 12", schedule 40 usually corresponds to standard weight, and schedule 80 is usually the same as extra strong, but there the similarities between the systems end. So, for example, a three-inch standard weight (schedule 40) pipe has a specified i.d. of 3.068", and an o.d. of 3.5". A three-inch extra strong (schedule 80) pipe also has a 3.5" o.d., but the i.d is only 2.9", because its wall thickness is greater. The reason that the o.d. must remain the same size for all strengths of a nominal size is that pipe is often joined using threaded couplers, which would be impractical if outside diameters varied in each case. The availability of pipe varies dramatically from size to size; certain strengths are not available in all sizes, and the wall thickness of a particular strength is different for larger pipes. With pipe, it is best not to assume what the actual dimensions will be; check with your supplier for exact specifications.

Although most pipe is made from carbon and stainless steel, it is also available in some sizes in brass. In most cases, pipe has a rough outer surface, since it is not really intended for decorative purposes. For utility scaffolding and railings, there are a number of versatile systems available that provide kits made of joints for assembling

Tubing.

lengths of pipe into rigid frames. The joints often use set screws (see Chapter 8) to allow for rapid assembly and disassembly of parts.

Unlike pipe, metal tubing always is measured by its outside dimensions (o.d.). In fabrication, tubing is used as an alternative to making things from solid bar, since it often provides sufficient strength for an application without the weight of a solid extrusion. Tubing is available in round, square, and rectangular profiles, in a wide variety of sizes and wall thicknesses. Specialized manufacturers produce unusual tube profiles for decorative purposes, although these products are substantially more expensive than standard tubing.

Pipe and tube are made using a variety of processes, all of which affect the appearance of the extrusion's outer surface. The two basic categories are welded and seamless. Welded pipe and tube is made by rolling flat stock into a tube, and continuously welding the material at the seam. Some extrusions produced this way are ground to remove or reduce the bump produced at the seam, but this is not always the case. Seamless pipe and tube is made in a number of ways, most involving feeding hot metal past a piercing die (that creates a hole) and a series of reducing mandrels. Tube produced this way is called hot-rolled, like sheet. For the most part, however, hot-rolled tubing does not have a rough surface like hot-rolled sheet products.

To produce tubing with a cleaner finish, the hot-rolled tube can be further reduced by cold drawing. Cold drawn tubing is hot-rolled tubing that has been pulled through a tapered mandrel that squeezes the tube to the desired diameter. Cold drawing is frequently used for thin-walled ($^1/_{16}$" or less), small diameter, and decorative tubing products in carbon and stainless steel. Just as there is a secondary level of companies that makes "after market" products from sheet, there are others that produce ornamental tubing in exotic shapes, with flutings, textures, and other details. Most decorative tubing is made from thin-walled material. There are also processes available to make tapering and bulging shapes from tubing (see Chapter 6).

Extrusions are the "skeletal" elements of fabrication, just as sheet functions like a skin or shell. Extrusions have a vocabulary of their own that anyone intending to design with metal should master.

CHAPTER 3

CUTTING AND SHEARING

From a practical standpoint, the taxonomy of metal cutting can be divided into two categories: methods for cutting shapes out of sheet and plate, and methods for cutting bar stock to desired lengths. Shape cutting can be further divided into non-linear cuts, linear cuts, and circle cutting.

SHAPE CUTTING IN SHEET AND PLATE

NON-LINEAR CUTS

Very thin or soft sheet metal can be cut by hand with high-strength shears called snips. Snips come in a variety of configurations to accommodate straight or curved cuts, as well as left- or right-handed cuts. For mild steel that is thinner than 20-gauge, snips can be very useful for cutting out simple shapes, such as ovals. Snips are also used to remove small ridges of metal, called burrs, that are left behind by other cutting methods. There are also air- and electric-powered snips, which rapidly open and close as the user guides the device along a cut.

The most basic hand tool used to cut thicker material is a hacksaw. Hacksaws have a band of high-strength steel, called a blade, that is suspended in tension on a hand-held frame. The blade has either serrated teeth or an edge coated with abrasive particles. The user pushes and pulls the blade against the metal piece with sufficient force to scrape away tiny particles of the metal, called chips. The friction created by this action also generates large amounts of heat, on both the blade and the part. Hacksaws are subject to frequent breakage; under the best of circumstances, they provide a cutting rate of about $1/2$" per minute in $1/4$"-thick steel plate. Curved cutting is limited by the depth of the blade, and straight cuts require considerable skill and patience. A cut made in metal is called a kerf, and depending on the process used, the kerf has different widths. With most toothed cutters, the kerf is equal to the thickness of the blade.

The electric jigsaw employs basically the same cutting process as the hacksaw, but to considerable advantage with an electric motor. The motor causes a short blade protected by a horizontal guard to oscillate rapidly in a vertical or circular pattern. Jigsaws can be purchased with

A hacksaw cutting a
part held in a vise.

Jigsaw.

a simple grip or a steerable knob to allow for precise control of curved cuts. Some models permit the attachment of spray coolant devices, which distribute a mist or liquid flow of water-based oil to keep the blade from overheating. Jigsaws are extremely useful for short production runs of cut shapes in carbon steel or softer metals, of ¹/₈" to ¹/₄" thickness. It is better to use a guide, called a pattern, rather than try to "freehand" a shape. A pattern is usually made of plywood, cut smaller than the desired shape, so that by holding the jigsaw guard against its edge, the cutting blade will be located at the perimeter of the shape. Jigsaws produce a very clean, burr-free cut, although owing to their hand-held use, fabricators will purposely overshoot their pattern slightly and return with a sanding tool to grind the edge of a piece to the exact shape required. Straight cuts always require a guide.

For material thicker than ¹/₄" or harder than carbon steel, several cutting systems exist. The oldest method is called oxygen-flame cutting. Oxygen-flame cutting was first developed at the beginning of the twentieth century to capitalize on methods developed to bottle pure oxygen in pressurized cylinders. In an oxygen-flame cutting device, a stream of compressed oxygen is ignited in a nozzle by a burning mixture of oxygen and a fuel gas (usually acetylene or propane). The resulting flame stream is sufficiently hot to melt the metal in its path. By moving the cutting nozzle forward, one creates a cut that can be as much as three feet deep depending on the size of the torch. The kerf is often very wide and messy. Torches are usually hand-guided, although they can be equipped with mechanical pattern guides (normally based on a pantograph), or computer numerical control (CNC) guidance systems (see Chapter 19).

Plasma-arc cutting is similar to oxygen-flame cutting, but instead of using a flame to ignite the cutting oxygen stream, an electric arc is passed through a compressed gas jet in the cutting nozzle. The arc heats the gas to the point where it undergoes a phase change, becoming plasma. Plasma is the fourth state of matter; it consists of superheated gas made up of ionized particles. The plasma is heated to temperatures as high as 25 thousand degrees Fahrenheit, and aimed at the material to be cut. As the plasma converts back into a gas, it transfers much of its heat to the cutting zone and produces a tightly defined melt. The melted metal is blown out of the bottom of the part as it is guided along a cutting path. The kerf produced varies depending on the thickness of the material. Water is sometimes used in the cutting process to shield the torch from the intense heat, as well as to constrict the diameter of the plasma stream. Water also muffles some of the noise produced by plasma-arc cutting. Some large plasma cutting machines include a tank for submerging the parts to be cut in a shallow layer of water, while smaller machines incorporate the water in the cutting nozzle.

In recent years, advancements in microcircuitry have allowed for the development of very small, portable plasma cutters that weigh

A plasma cutter
mounted on an
automated rail designed
by the fabricator for a
specific job.

as little as forty pounds (older models weighed several hundred pounds). Most small plasma cutting machines employ a hand-guided torch, which uses ordinary compressed air for gas. The portable machines are designed for cutting carbon steel up to $3/4$" thick, which makes them very versatile.

While oxygen-flame and plasma cutting are relatively fast and inexpensive metal cutting techniques, the kerf they produce requires some cleanup in thinner stock, and very substantial cleanup in thicker parts. Two techniques exist for cutting thick metal very cleanly and accurately: abrasive water-jet cutting and laser cutting. Abrasive water-jet cutting, also called hydroabrasive cutting, can be used on many materials, including stone, glass, and plastics. A jet of highly pressurized water is forced into a nozzle and mixed with a stream of solid abrasive particles. The water and abrasive material mixture is then fired through a tiny nozzle in a highly focused stream, causing the rapid erosion of the metal in its path. The cut is usually controlled by CNC guidance (see Chapter 19). The maximum depth of cut for most hydroabrasive machines in carbon steel is $3/4$", and the kerf is very small (usually around one tenth of an inch). The speed of cutting is inversely related to the thickness and hardness of the material.

Lasers are also employed to cut metal as well as many other materials. The most common lasers used for cutting are called CO_2 lasers. A CO_2 laser consists of a high-intensity beam of infrared light that is focused with a lens into an even more intense beam. The focused beam is aimed at the desired cut point in combination with a jet of highly pressurized gas (carbon dioxide). As the laser melts the metal, the gas blows the liquefied material from the kerf. The kerf produced by laser cutting is very small and even (usually around one hundredth of an inch). The gas can also be focused slightly ahead of the laser, to preheat the surface of the metal to be cut. As with water-jet cutting, lasers are usually computer guided. This means that whatever shape you intend to have cut must be described in a computer file that the laser 's controlling computer

Water jet cutting. A high pressure stream of water (a) is mixed with abrasive particles (b) in a mixing cup (c) and aimed at the cut zone, where the gritty water cuts by erosion.

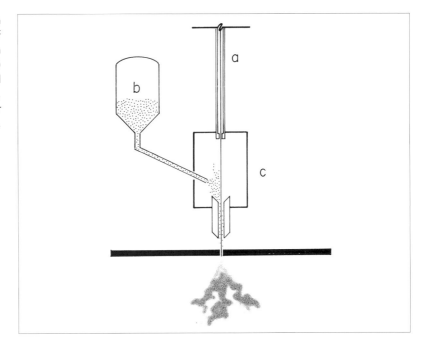

A water jet cutter. The particle board under the metal parts is used to make practice cuts to check the program.

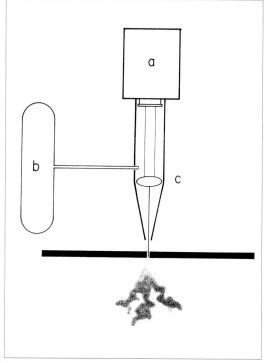

Laser cutter.

Laser cutting. A laser beam emitted by a laser (a) is focused to an even greater intensity by a lens (c). Carbon dioxide (b) is added under pressure to blow molten material from the cut zone.

Parts cut from sheet on a laser cutter.

can read. The setup time associated with both laser and water-jet cutting makes it sensible to use the processes only for larger runs of pieces so that these costs can be amortized. For highly complex individual parts like signage, however, both processes are excellent and cost-effective choices when compared with manual cutting.

LINEAR CUTS IN SHEET AND PLATE

The most common method for making straight, accurate cuts in metal sheet and plate is with a shear. A shear is a mechanically or hydraulically controlled cutting device that generates tremendous pressure at a highly localized spot on or along a straight line. (In the case of a spot, the process is called punching and will be described

Large shear.

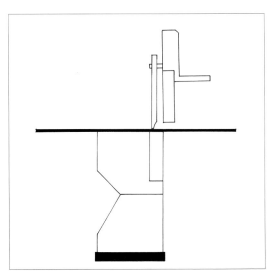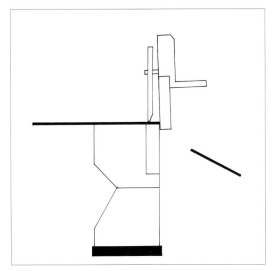

A hydraulic shear. The sheet is inserted between hardened cutting blades, one mobile, one stationary. The mobile blade is angled slightly off vertical, so that it is directly aligned with the bottom blade when it enters the sheet.

in Chapter 4.) Shears distribute their force along a high-strength steel cutting blade and a flat bed. The material to be cut is rested on the bed and pushed through a small gap between the bed and the cutting blade. The blade drops along the bed, which has a hardened edge separated by as little as .002". In this zone, the cut occurs using exactly the same mechanism that is employed in cutting fabric with a pair of scissors. No cleanup is required, although in some cases the cut may be sharp enough to require sanding. Shearing thick metal (over $1/4$" in steel) can produce slight deformations in the leading edge of the cut.

CIRCLE CUTTING

Circles can be produced in a wide variety of ways. Small circles can be punched from sheet using punched discs (see Chapter 4), which are easily obtainable in a wide variety of sizes. The only problem with punched circles is that they have a slight edge and surface deformation that might make them unsuitable for some applications. Circles can also be cut as slices from round bar on the machines described later in this chapter. Circles of all sizes can be plasma, laser, or water jet cut. They can even be cut with jigsaws using revolving tables, or by following round patterns.

A very specialized circle cutter exists for making circles in material up to $1/4$" in thickness. Called a circle cutter or a ring shear, the machine is basically a very large vise, with two circular cutting teeth that cut a circle from sheet or plate as they spin past it. In effect, this tool looks and acts like a giant can opener. The metal to be cut

Circle cutting machine.

is secured at its center, and the desired radius is set by positioning the the cutting wheels at the perimeter of the circle. The metal plate is then spun, and the pressure exerted by the teeth is slowly increased until the entire edge of the circle is cut. Circle cutters are frequently found in metal spinning facilities (see Chapter 6), but the common need for circles makes them useful for many fabricators.

CUTS IN BAR STOCK AND OTHER EXTRUSIONS

All of the methods previously described in this section for shape cutting can be used for cutting extruded materials, but none are commonly employed, with the exception of the hacksaw and specialized versions of a shear made for bar stock. A shear called a notcher is specifically made for fast cuts on angle. Notchers have a V-shaped shoe that holds angle stock and a cutting die that shears a 45-degree notch into one side of the extrusion. By making four cuts in a single angle bar, a fabricator can then bend the angle at the apex of the notches to quickly produce a "box" frame.

For other cuts in extrusions, most fabricators rely on abrasive chop saws. Chop saws use a resin wheel impregnated with abrasive grit which is rotated at a high speed to generate a cut in the metal. The saw can be adjusted to make angled cuts, usually up to 45 degrees. Chop saws are relatively inexpensive, but they have several drawbacks in service. First, it is difficult to make precise, repetitive cuts. The cut itself is rough and requires substantial cleanup. Further, the cutting discs wear down quickly, particularly on harder materials like stainless steel.

A chop saw cutting extrusions.

A much more expensive cutting machine called a cold saw delivers a far superior cut on extrusions. Cold saws are about the same size as chop saws, and they are usually limited to cutting extrusions up to 6 inches in diameter. A cold saw uses a high-strength steel blade with serrated carbide teeth and is turned by a slow motor (usually 30 or 60 r.p.m.). A pump attached to the saw pours coolant on the blade that drips into an attached basin for recirculation. The extrusion to be cut is held in a vise under the blade, and like the chop saw, the blade can be positioned to create an angled cut. Special cold saws are available that can rotate up to 90 degrees for splitting or slitting applications. Cold saws are ideal for making precise, clean cuts on extrusions. They perform particularly well on tubing, and the cut they produce needs little or no cleanup.

A cold saw cutting an extrusion.

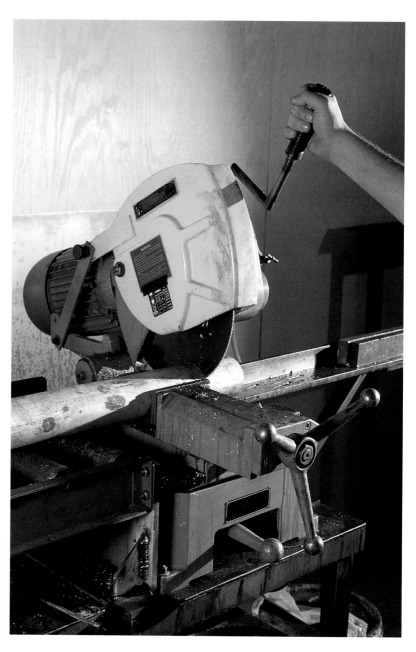

A miter cut on tubing prior to welding.

Band saws are often used to cut extrusions with dimensions larger than 6 inches, or bundles of many pieces clamped together. The saws are referred to as horizontal or vertical, depending on their configuration. Band saws cut using a serrated metal band that circulates slowly to prevent snagging. Horizontal band saws are either hinged or move along a bar guide, and the rate of blade travel is usually controlled by a hydraulic piston. Like the cold saw, band saws use a coolant pump and collection trough. Some production band saws are equipped with computer operated hydraulic vises that automatically feed new stock into the blade for large automatic production runs. Band saws are commonly found with cutting capacities as large as 24 inches.

A large horizontal band saw.

DRILLING AND PUNCHING

A very large portion of the work performed in metal fabrication involves placing holes in parts. For very precise holes with tolerances of .01" or greater, the appropriate machines to use are either the mill or the lathe (see Chapter 7). The majority of holes do not need to be so precise, however. Often, as with holes used to join two parts using a screw, it is desirable to have a substantial amount of tolerance in the hole to account for the vagaries and slight misalignments caused by unexpected field conditions. For these holes, there are two basic strategies. The first is to drill the hole using a hand drill or a drill press equipped with a drill bit. The second strategy is to use a punch equipped with a punching die.

DRILLING

Drills are used to produce circular holes, usually in the range between $1/16$" to $1\text{-}1/4$" in diameter, in a wide variety of metals and stock thicknesses. Please note that we say circular, rather than round; the holes created with a drill are not entirely symmetrical. This is because drill bits—the cutting tool used in drills to create holes—are subject to vibration, or wobble, which slightly deforms the hole shape. Also, any dullness in the bit will cause stock to be removed unevenly. This usually does not affect the performance of a drilled hole since the hole is not meant to be perfect. Holes that need to be precise should be milled. Drilled holes can be made fairly precise by using a tool called a reamer. A reamer is a cutting tool that is placed in a drilled hole which has been slightly undersized. The reamer is turned in the hole, often by hand, and it scrapes away a thin layer of rough metal at the periphery of the hole, opening the hole to the desired size. For very precise hole conditioning a special grinding tool called a hone is used. Another device called a broach can be used to cut drilled holes into different shapes, such as square and hexagonal. A broach is a long shaft with a series of cutting teeth that gradually taper from round to the desired shape. The broach is inserted into the hole and, using hydraulic pressure, pushed through to cut a new profile.

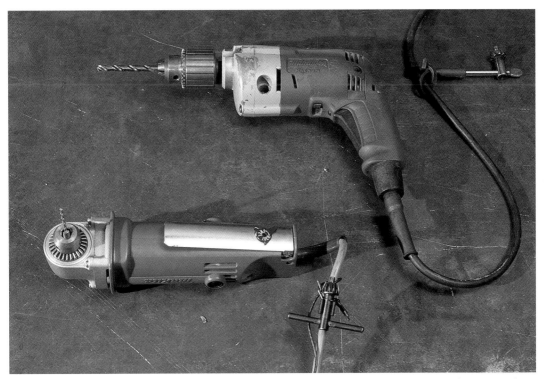

Hand drills. The one on the bottom is an angle drill, used for drilling in very tight spots where a regular drill would not fit.

A reamer is used to open and smooth rough holes made by drilling.

A broach is inserted into a round hole and pushed using hydraulic pressure. The cross section of the broach gradually shifts from round to square, so that as it passes through, the hole is made square.

Drill bits used for metal are usually milled from solid rods of the intended hole diameter. A sharp-edged double spiral is cut deeply into the rod. When the bit is spun by a drill, the cutting edge peels away layers of metal from the workpiece. These layers are called chips. The grooves in the bit have the secondary effects of creating a channel for the chips to follow out of the hole and allowing cooling oil to reach the cut zone. Drill bits are made from or coated with materials that are harder than the material they are meant to cut. Typical materials used include black oxide treated steel, cobalt, titanium nitride, and carbide. Bit tips are cut at a 118-degree angle for typical shop applications; bits cut at a steeper 135-degree angle are used for harder metals.

Usually, the drill operator will hammer a small divot or drill a very small diameter "pilot" hole at the center point of the hole location. Drill bits tend to "walk" if they are fed directly into material without a pilot hole. Holes larger than about $^3/_8$" are usually drilled in steps; first a pilot hole, then a $^1/_4$" diameter hole, then the $^3/_8$" hole. This is because the amount of stock removed from a larger hole generates tremendous frictional heat that quickly destroys bits. It is better to save large bits for cutting on the perimeter of a hole only.

Drill bits are usually 3" to 6" long; their length limits the maximum material depth through which they can travel. The typical drills used in fabrication shops are called jobber's drills. They are kept in indexed sets, in sizes from $^1/_{16}$" to $^1/_2$", in increments of $^1/_{64}$". They can also be found indexed by a code letter (ranging from from A to Z), which matches the drill to an appropriate tap size (see Chapter 9). Drills larger than $^1/_2$" are usually made with a $^1/_2$" diameter shank, so they can be fitted into standard drill chucks. Although drill bits are made up to 2" in diameter, the larger sizes are expensive, wear out quickly, and require many steps from pilot hole to the final diameter. Another cutter, called a hole saw, often makes more sense for large holes, especially on materials of $^1/_2$" thickness or less. Hole saws are cylindrical cups with cutting teeth arranged around their diameter. They are attached at their center to an arbor, which has a shaft for insertion into the chuck, and a drill bit (usually $^1/_4$") to hold the center of the hole. Hole saws with diameters of about 1" to 6" are most effective in metal. Special carbide-tipped hole saws can be used effectively in stainless steel and other hard metals. Another device, called a fly cutter, uses a single adjustable cutting tool to trace the desired diameter. Fly cutters can be used on drill presses, but they are probably more effective on mills and lathes, which are built solidly to prevent the kind of wobbling found even on very large drill presses.

There are several other kinds of bits commonly used with drilling equipment. These include countersinks, which are bell-shaped cutters used to make tapering holes that match flat-headed screws that are mounted flush into the surface of parts. Counter bores are used to create stepped holes, which provide seats for fillister head screws. Chucking reamers serve the same purpose as hand reamers, but they are spun on a drill press. Taps can also be used on drill presses to create threaded holes, but in practice they are employed when tapping clutch attachments are also used (see Chapter 9).

The equipment used to spin drill bits varies in size from handheld electric and air powered drills to the larger drill press. All drills have a spinning shaft, called a spindle, which is attached to a vise-like device called a chuck. Chucks have a set of three adjustable jaws that tightly clamp against the smooth shaft of the bit. Most chucks open to accept a maximum $1/2"$ diameter shaft. Some chucks

The tools used to drill holes. Top row, from left: hole saws, including one mounted to a hole saw arbor; countersink bits; drill bits. Bottom row: carbide-tipped circle cutters, for larger holes in hard metals like stainless; an adjustable circle cutter (called a fly cutter); and a diameter gauge for drill bits.

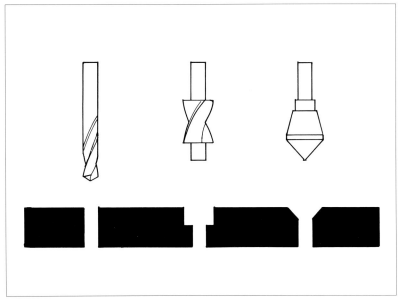

Left to right: drill bit, counterbore, countersink bit.

are opened and closed using a chuck key, which binds into holes and teeth along the perimeter of the chuck. Other chucks, called keyless chucks, are opened and closed by twisting the chuck by hand. Keyless chucks are most common on hand drills.

Hand drills are most effective for drilling holes up to about $^1/_4$" diameter. Their advantage is portability, but their disadvantage is that they are unlikely to produce holes that are perpendicular to the surface of the drilled material, and they are subject to wobbling. The accuracy of hand drills can be greatly improved by attaching them to a jig, which controls the angle of drilling and the feed rate. Some hand drill jigs have magnetic bases that make them very stable.

Drill presses are machines that combine a drill mounted in a sturdy head with a table that is machined to be flat and precisely

A portable jig used for holding a drill parallel on plate where a drill press cannot be positioned. Better jigs have magnetic bases to keep the drill steady.

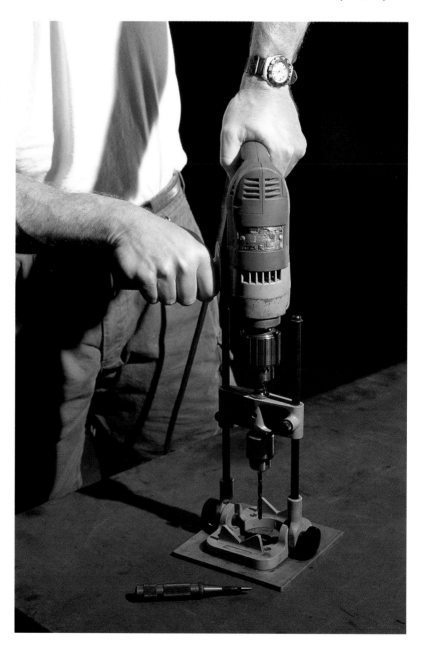

perpendicular to the chuck. Bench-mounted drill presses have
tables that can be adjusted anywhere from 18" to 30" away from
the chuck. Floor-mounted units allow greater adjustment (usually
from 39" to 49"), so very large work pieces can be placed beneath
them. The chuck on drill presses is usually 14" to 20" from the
column that supports it. That means that a hole can be placed at
mid span on a 28"- to 40"-wide circle. For hole drilling on wider
pieces, it is necessary to use a hand drill.

The table on drill presses provides a sturdy, level base for securing
work pieces beneath the chuck. Pieces are either clamped directly
to the table or placed in a vise that is itself clamped to the table.
Cooling oil is applied to the cut either by hand with a brush or with
a portable coolant misting system that sprays a fine mist of oil through
an adjustable nozzle to the drilling zone.

Most drill presses are variable speed and have a hand-operated feed. The bit is lowered (fed) to the work piece by a hand-operated spindle that transmits the sensation of the bit cutting the material to the operator. This kind of drill press is called "sensitive," as it allows the operator to make fine adjustments to the feed according to the "feel" of the bit as it cuts. Larger drill presses sometimes have multiple spindles so the operator needn't stop to put a new bit in a chuck.

PUNCHING

Punching is the other main strategy used to create holes in metal parts. A punch is a tool that applies tremendous force to a very small area, and in doing so causes the material to shear away. Punches focus force into a hardened metal cutter (called the punch), which is made to the shape of the desired hole. The punch pushes through the work into a void that is the same shape as the punch, called a die. Punches can use mechanical advantage to generate force, but larger punches usually rely on hydraulic pumps. Punches vary in scale from small, hand-operated units that can create small holes in sheet metal all the way to very powerful automated punches that can cut holes with diameters as large as 12" in ½"-thick plate.

One of the greatest advantages of punching over drilling is that punches can assume many different shapes; square, oval, rectangular,

Hand-operated punch.

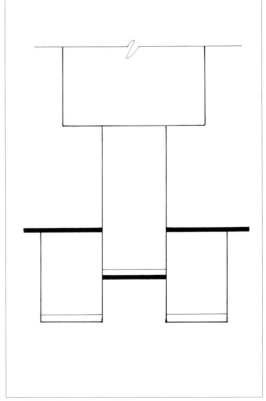

Punching. A punch is forced through a plate into a matched die to create a hole.

and pill-shaped punches are very commonly stocked. Custom punches of irregular shapes can also be made. This means that punching can compete with many cutting systems including laser cutting, depending on the exact nature of the piece and the size of the production run. The disadvantage of punching is the limited range of thicknesses that even very large punches can handle. However, a very large percentage of all fabrication is performed on material less than $1/2$" thick, so punching is frequently a desirable option for larger runs.

A punched hole can be designed to be very accurate. One slight detraction from many punched holes, however, is the surface deformation that can occur near the point of shearing. This deformation is most prominent on thin sheet or in large holes in thicker plate. The setup time for a punch also adds cost, especially on very short runs. Care must be taken to insure that the punch and die align precisely. Otherwise, the punch will be damaged as it enters the die. This is especially important on punches with sharp corners, such as squares.

There is a category of very large machine tools simply called "fabrication centers" that combine heavy duty groupings of punches with plasma cutters mounted on computer-controlled guides over large worktables. Fabrication centers process flat plate into finished components without the necessity of moving a piece from one workstation to another for cutting and hole drilling. Some fabrication centers even have trap doors that automatically release small finished parts as they are completed. Fabrication centers are used for runs of heavy equipment and machinery, such as tractor frames. Another machine called a rotary punch is used in large runs of parts with multiple holes, such as registers and grilles. The sheet is fed between a pair of multiple punches and dies that are mounted to drums. The drums rapidly turn as the sheet progresses between them, producing rows of evenly stamped holes.

BENDING AND ROLLING

Bending and rolling are the processes used to curve and fold flat sheets and straight extrusions. Both rely upon the ductility of metal, which is a measure of a material's ability to stretch and deform without breaking. The technical term used to quantify this characteristic is "the modulus of elasticity," symbolized by the lowercase letter "e" in structural formulas. All solid materials pass through three phases as increasing amounts of force are applied to them. The first phase is the elastic phase, when a material will spring back to its original shape when a load is removed from it. The second phase is called the plastic phase, when the material deforms under force and retains the deformation even after the force is removed. This is the state that is usually desired in metalworking. The third phase is the point of rupture, when the material not only deforms but breaks under a heavy load. This phase is usually avoided.

Bending and rolling are accomplished by applying a pressure and a counter pressure to a metal object. To control the shape of the bend, the piece is held firmly against a third point, called a fulcrum. A fulcrum, also called a die, is a piece that is the same shape as the desired bend, so when the forces are applied, the metal object conforms to the fulcrum's profile. If you imagine holding a length of thin rod in both of your hands, setting the rod against your thigh, and pulling down on either end of the rod, you have envisioned the principle of mechanical bending. One hand provides pressure, the other hand provides counter pressure, and your thigh serves as a fulcrum. If you bent the same rod against something with a narrower radius than your thigh—say, your ankle—you would either get a tighter bend, or the rod might even break if you exceeded its modulus of elasticity.

You will notice that while it is fairly easy to bend long metal rods by hand, shorter rods of the same diameter are less easy to bend. This is because on the long rod you apply force at a greater distance from the point of bending than on a short rod. You have increased leverage, which is also known as mechanical advantage. The tendency of a force to cause rotation (or bending) on an object is called moment. When the distance of a force from the

An interesting jig for bending flat bar.

point of rotation is increased, the magnitude of the moment increases, even though the force stays the same. Even people who never work with machinery will apply this rule instinctively; the principle is summed up by the classic example of a heavier person held aloft by a lighter person sitting further back on the opposite side of a seesaw.

BENDING EXTRUSIONS

Most devices used to bend metal rely on mechanical advantage. The basic tool used to bend extrusions is called (not surprisingly) a bender. A bender uses a steel pin swung by a long lever arm to push the work piece against a hardened steel fulcrum. The tool is used to create both gentle curves and very sharp, straight bends on a wide variety of stock. Benders are handy for making bends in solid bar and structural extrusions, as well as thick-walled tubing and pipe. The leverage of the long arm usually provides enough moment to generate bends; for larger work pieces, a hydraulic arm is sometimes added to provide additional force.

When a bend is too tight, the work piece will crack. One strategy to overcome this problem is to heat the piece, bringing it close to a plastic phase. Bends formed on hot material can usually be much tighter than unheated, or cold-formed bends, but care must be taken to prevent unwanted changes in the strength of the material (see Chapter 1). The metal extrusion can be heated with a torch or placed in an oven prior to bending.

While a bender can certainly be used to create radius bends in hollow pipe or tube, thinner walled tubing will often deform slightly in cross section, or even partially collapse at the bend point. Historically, accurate tube making and bending was one of the more vexing technological problems of the nineteenth century. Many mechanical applications require that the cross section of tubes carrying gases or liquids remain constant. From an aesthetic standpoint, even very slight deformations in cross section at bend points are visually obvious, and therefore undesirable. Clean bends

A hand-operated bender for solid bar and rough bends on pipe and tube.

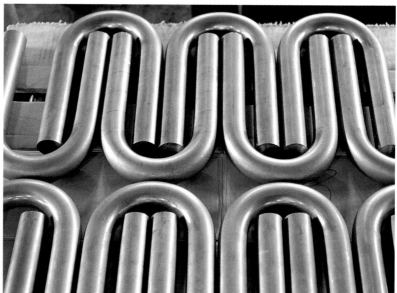

Tight bends on tube.

can be made in thin-wall tube either by compressing a resistant material, such as sand, into the tube prior to bending, or by inserting a flexible "mandrel." A tube bending mandrel is a string of short, solid pieces, each the size of the tube's inside diameter, which can conform to a curve like a string of beads. Sand- or mandrel-packed tubing is formed by forcing it against a fulcrum with one or more hydraulic pistons. The tube behaves like a solid extrusion during the bending process, retaining its cross section with minimal deformation. Once the bend has been made, the mandrel or sand is rammed out of the tube.

Bending a tube by passing it through a set of mandrels.

Tube-bending mandrels.

The basic tool used to bend flat sheet is called a hand brake. A hand brake is a tool with a long, hardened, horizontal steel bed and an adjustable panel running above the length of the bed, called a nose bar. The nose bar has a set of forming dies, called teeth, mounted with screws onto its face. The teeth, which serve as the fulcrum in bending sheet, have pointed edges set opposite the bed. Usually the teeth used are identical, although groups of them can be removed to create partial bends along the edge of a sheet. They can also be raised or lowered in relation to the bed, to make a tighter or wider radius bend on the sheet. The bed is hinged so that one side can rotate toward the teeth. The rotating side is equipped with a pair of levers, usually fitted with counterweights, to provide the operator with mechanical advantage. The work piece is slid into the gap between the teeth and the bed. The levers

A brake. The plate to be bent is inserted between the bed and the fingers on the nosebar (top) and the plate is lifted against the fingers to create a bend (bottom).

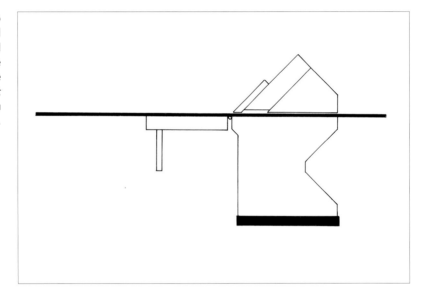

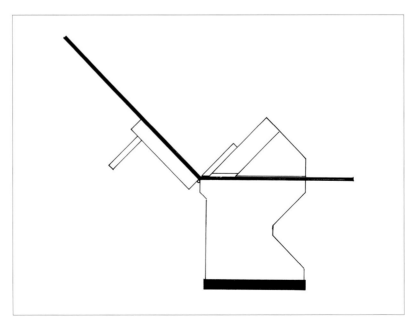

are lifted by hand, which causes the bed to rotate and push the work' piece against the teeth, thereby creating a bend. Hand brakes usually will bend mild steel up to $1/8$" thick, and stainless steel up to 16-gauge. Thicker material must be bent on brakes powered by hydraulic force.

Hydraulically powered brakes, called press brakes, are configured differently from hand brakes. They usually consist of a long male or female die secured to a sturdy base. A matching die is fastened above the lower die and rammed into it while a sheet is stationed between them. This machine allows for much more complex shapes and sequences of bends to be created in all thicknesses of sheet. We will discuss the additional uses of a press brake in Chapter 6.

ROLLING SHEET AND EXTRUSION

Rolling describes the process used to add wide, continuous curves to sheets and extrusions. Rolling is similar to bending in the use of pressure, counter pressure, and a fulcrum. Unlike bending, however, a rolled work piece is fed in motion through spinning stationary dies. A sheet roller, also called a slip roller, consists of a set of two or three cylindrical dies set parallel to each other. One end of the sheet is inserted between the dies, which clamp the sheet tightly. The dies are then spun, and friction pulls the metal through the rollers. As the sheet passes through, it is forced against the dies and assumes a curved profile. The distance between the dies is adjusted to create the desired radius in the sheet being rolled. The slip roller can either be hand cranked or electrically operated. Some rollers for sheet are embossed with patterns that are imprinted onto the sheet.

Extrusions are run through sets of dies that look a bit like wheel hubs on automobiles. The dies are stout, round discs with the shape of the extrusion to be rolled formed into their edge profile. Tighter rolls may require several passes, each time reducing the distance between the dies slightly. With extrusions, cross sectional deformation is avoided by using rammed sand or a mandrel, just as with bending.

Rolling a plate through a set of adjustable rollers. By changing the configuration of the rollers, different radii are produced.

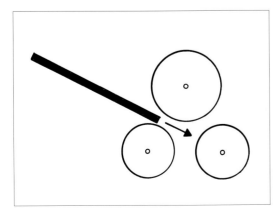
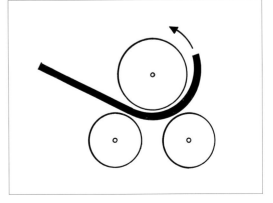

Rolling thick plate into a circle. The top roller is removable so that the finished part can be taken off.

Material can be rolled flat, into a spiral coil, or even into a conical section depending on the skill of the roller. When material is rolled in the direction that it most naturally will bend, the bending is described as being in the "easy" way. When material is bent in the direction of greatest resistance, this bend is referred to as the "hard" way.

Sometimes with a rolled extrusion, the material on either end of the bar cannot receive the roll. This is because the dies are spaced far enough apart that as the last of the material passes through, it is not held by both a pressure and a counter pressure. The amount of unrolled material varies, but a good rule of thumb is to expect to lose about one foot from either end of a length.

Rolling an I-beam
the easy way.

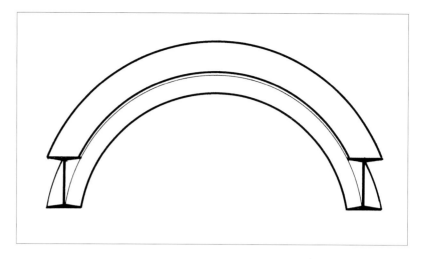

Rolling an I-beam
the hard way.

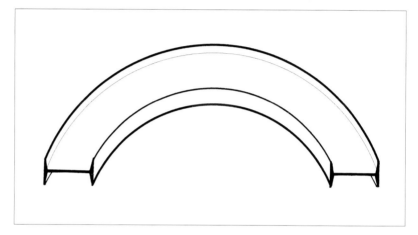

Rolling on very large
structural channels

PRESS WORK AND SPINNING

PRESS WORK

Press work is a group of processes that include stamping, deep drawing, fine blanking, embossing, forging, coining, and upsetting, all of which create parts from sheet metal using tremendous forces generated by hydraulic rams. Stamping generally refers to the process of adding holes and some three-dimensional contours to flat sheet pieces, called blanks. The blanks have usually been cut to assume the two-dimensional profile needed for stamping. Unlike punching, a number of holes of various size and shape and different contours are embossed into the surface of the blank in a single step. Deep drawing is the process used to make fully three-dimensional parts, like canisters and artillery shells, from

Press brake.

flat blanks. Fine blanking is the process of producing flat parts with complex profiles from thin, flat sheet using mated dies, much like in punching. The difference between punching and fine blanking, however, is that the part knocked out of the sheet with fine blanking is the finished part, while the part removed in punching is waste. Embossing is the application of a subtle pattern on the surface of a metal sheet. Forging is the process of forcing a metal blank to assume the shape of a die placed beneath it by subjecting it to one or more powerful blows. In coining, the same process is applied, but the pressure is applied slowly and continuously. Upsetting increases the diameter of a part by striking it with a heavy blow and forcing it to expand into a surrounding die. Bolt and screw blanks are often made this way by forcing a solid rod to mushroom at one end to create a screw head.

Press work, a sort of self-contained world in metal fabrication, has its own strategies for solving a variety of detailing conditions on sheet metal parts. Flat parts are cut, punched, embossed, bent, and even rolled (called curling in press work) by using different kinds of dies on a hydraulic ram. Most press work performed by job shops is for small parts, although the same processes are used in mass production for large, complex components, such as automotive parts. The setup costs for large parts can be in the millions of dollars, however, and this kind of work is well beyond our scope.

Press brake dies.

Hydraulic ram.

A press brake. A die (a) is forced by a hydraulic ram against a matching shoe (b), between which the sheet to be bent is placed.

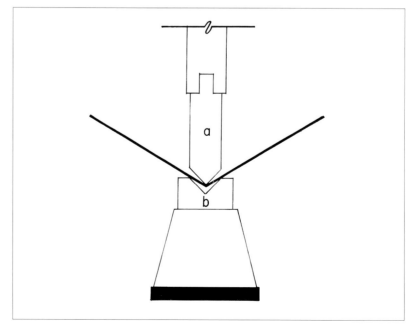

There are many different ways to employ the ram to make parts from sheet. In deep drawing, a female die is pushed against a blank, forcing it to stretch over a stationary male punch. Deep drawing is good for producing fully three-dimensional parts, and it is often a competitor for another process called metal spinning that we will discuss later in this chapter. The advantage of deep drawing is that it can be used to make much more complex shapes than those produced by spinning. Stamped parts are usually made by forcing a hydraulic ram against one or more punches that push into the blank and force it into a stationary die block, which contains a void that is the female opposite of the punch. The punches either form or cut, depending on what is needed for the part. Bending dies can be used alone or in sets to produce one or a series of straight bends in sheet, as described in Chapter 5.

There are many technical considerations in press work that require a substantial amount of both technical knowledge and practical experience to evaluate. The only person who is really qualified to make suggestions about parts is the operator of a press work shop. It is very important to bring the fabricator into the design process as early as possible to guide decisions about parts. A few commonsense rules are a good starting point for exploring possible designs:

- Sharp bends in the contours of a formed piece can cause tearing. Dramatic level changes are also problematic.

- Sudden changes in the shape of a flat blank can cause problems by creating points of weakness that might fail when they are punched or fine blanked (think of an hourglass shape).

Deep drawing dies.

Deep drawing.

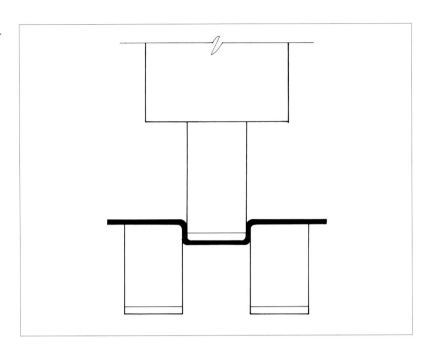

Press work parts. The "hat" is a good example of a part that could only be made using deep drawing (and not spinning) since it is not symmetrical.

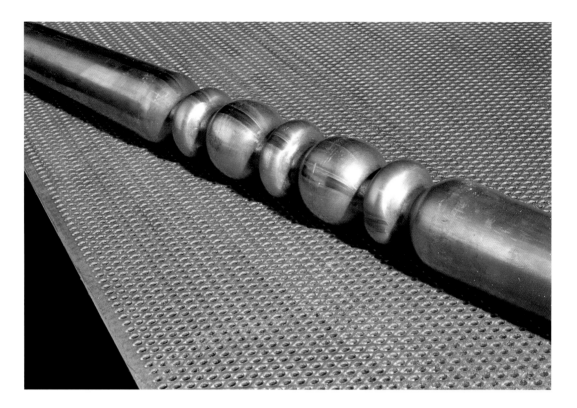

A part made from tubing formed by necking and bulging.

- Parts cannot have undercuts, otherwise they can not be removed from the die after forming.

- Metal parts will sometimes 'spring back' after forming. Dies may need to be designed to compensate for this so that parts actually 'spring back' into what is the desired position.

Other processes related to press work include necking and bulging. Necking and bulging are used to narrow and expand the profile of tubular blanks. In necking, the tube is forced by a ram into a narrow die, causing the tube to develop a taper. In bulging, an expandable, bladder-like pouch filled with liquid is placed inside the tube and pressed by the ram. A die surrounds the outside of the tube, and as the tube is pushed by the pouch it deforms to match the profile of the die.

SPINNING

Spinning is another process for generating complex curves from flat sheet. In the case of spinning, the process requires that the shapes created be radially symmetrical, like half spheres or nose cones. Spinning is commonly used to make parts such as lamp shades and bases, bowls, nose cones for missiles and planes, and numerous decorative items.

Spun shapes begin as circular, flat sheets, called blanks, that are usually made from thin material. Circle-cutting machines, like those described in Chapter 3, are a common sight at spinning facilities. The blank is lubricated and secured at its center on a lathe adjacent to a chuck. In this application, the chuck is a wooden or metal master form, which is the exact shape of the desired piece. As the lathe spins the blank, it is gently formed against the chuck by a variety of stick-shaped forming tools. A properly formed part should be of uniform thickness after completion. Spun parts tend to have a series of concentric lines called "grain" perpendicular to the axis of the lathe.

Spinning facility.

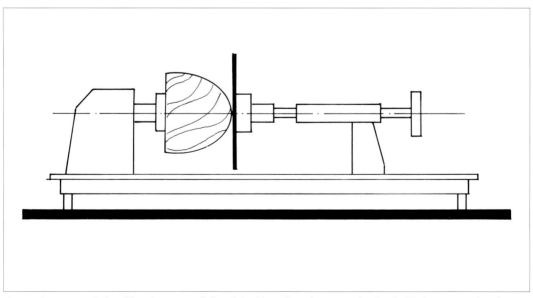

A metal spinning lathe. The sheet metal disc (black) is placed against the buck. Tools are used to form the metal against the buck and assume its shape.

Spun parts.

The forming tools used in the spinning process are made from wood and polished metals and are held in long handles so that the operators can push firmly against the blank. The tools are either blunt-tipped, sharp, or diamond-pointed (for trimming), or equipped with a wheel to create a roll-over or bead at the edge of the piece. The spun part can be trimmed, polished, beaded, or even knurled while still attached to the lathe.

Spinning is suitable for thin sheet only; good rules of thumb for maximum thickness are 14-gauge for steel, 20-gauge for stainless, about $1/8$" for aluminum, and $1/16$" for copper, brass, or bronze. The structural characteristics of domed shapes make spun parts much more rigid than they would be in their flat state.

Tools used for metal spinning.

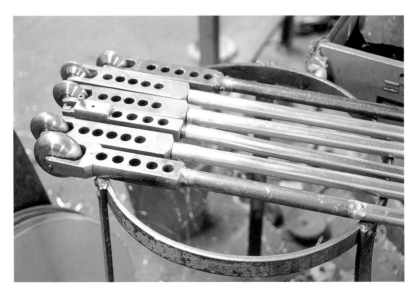

As with stamped parts, it must be possible to remove the finished piece from the chuck when it is completed. For this reason, most spun shapes are relatively simple, growing more narrow toward the front like a cone. More complex shapes with undercuts are made possible by using collapsible chucks, which can be taken out of the finished piece in sections. Another trick is to place a smaller chuck off center so that it generates a shape larger than its own diameter, much as a disc spun off axis will trace a larger circle. A third strategy is to use several chucks for the same piece—one for the overall shape, and then a smaller one for a "necked" (more narrow) detail.

Custom chucks are expensive to produce, although they are often less expensive than the forms needed for press work. Metal spinning companies tend to keep large libraries of standard shapes and unusual chucks that they have produced over the years. It is often possible to reuse an existing chuck (assuming that it isn't proprietary) instead of investing in a custom one.

A metal chuck used for spinning on a lathe.

CHAPTER 7

MILL AND LATHE TECHNIQUES

The mill and the lathe are the fundamental devices that make the construction of machinery possible. Mills and lathes are therefore defined as "machine tools," and their use is called "machining." Although relatively crude wood-cutting lathes have existed for at least 5,000 years, the original metal-cutting lathe produced by Henry Maudslay around 1800 made accurate and repeatable screw cutting possible for the first time. Accurate threads with even size and spacing are required for predictable power and motion transmission (see Chapter 17). The subsequent invention of the first true mill by Eli Whitney in approximately 1818 was one of a handful of pivotal events that propelled the Industrial Revolution. It is not inconceivable that a generative history of every machine in the world today could be traced back to a few prototype mills and lathes built two hundred years ago.

Machining is one of the most expensive metal fabrication processes owing to the substantial setup time and expertise required of machine tool operators. Machined parts should be specified only when their primary attribute—precision—is required.

THE MILL

A mill is defined as a tool that uses rotary cutters to remove stock from a work piece that is clamped to an adjustable table. The table can be smoothly adjusted by the operator in the x, y, and z directions relative to the cutter. This allows the cutter to follow an infinite number of paths as the work piece is fed against it. The cutter itself can also be moved up and down like a drill bit on a drill press. Above the mill table is a unit called the milling head, or turret. The turret contains the electric motor and sets of gears or pulleys that turn a spindle. Cutters are held securely by the spindle. The type of cutter most frequently used in milling is called an end mill. End mills have serrated teeth on their edges and bottoms for removing material axially and radially.

There are two primary categories of milling machines: the vertical and the horizontal. The most common mill in smaller job shops is called a vertical turret mill. The turret of this machine can be rotated

Mill cutters.

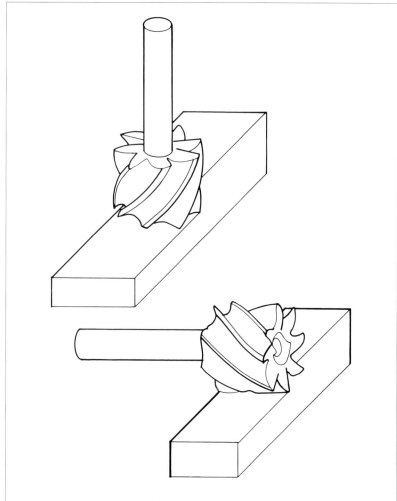

How work is fed into
a cutter on a mill.

Vertical mill.

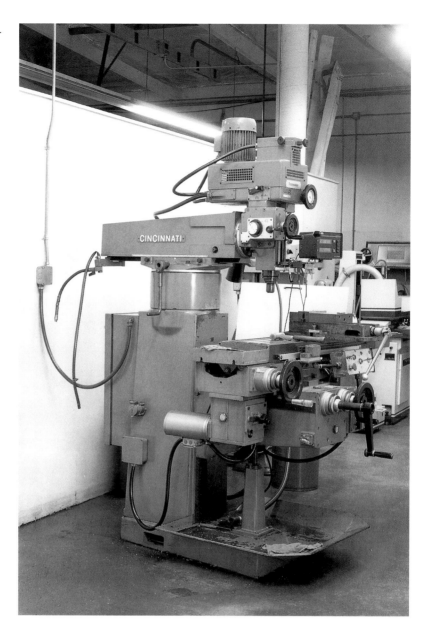

so that the cutter may engage the work at any angle, although usually it is held at 90 degrees to the table, like a drill press. The spindle is held in a quill, which can be raised or lowered toward the table by means of a crank or lever. The crank is for precise or fine motion, the lever for coarse motion.

Horizontal milling machines have the turret mounted parallel to the table rather than perpendicular. The two types of horizontal mills are plain and universal. On a plain horizontal mill the turret is fixed. On a universal horizontal mill the turret can be swiveled. Swiveling allows for the user to cut spiral or helical grooves, such as those on gears or threaded fasteners.

Many of the operations performed on a mill can be accomplished using more crude tools, but what distinguishes the mill is the precision that can be achieved with it. The dimensional accuracy required by moving parts simply cannot be attained consistently without one.

The mill table motion controls, called feed screws, are calibrated to adjust the position of the table to within a thousandth of an inch or less. The feed screws are controlled either manually by turning a hand crank or automatically by servomotors linked to computer numerical control, called CNC for short (see Chapter 19). It is possible to create extremely complex cutting paths by adjusting the feed screws simultaneously, but only highly skilled machinists can perform such a task on a manually controlled machine. Prior to the introduction of CNC, special mills equipped with pantographs were built to replicate complex shapes by following a pattern contour. The computer has dramatically altered machining as mills are ideally suited for the application of CNC.

The speed at which the cutter spins can also be controlled. This is accomplished by adjusting the position of the power transmission belt on two sets of pulleys of graduated diameter. Some mills have a finite set of possible speeds, while others have infinite variety within a specific range. Different cutting profiles, material hardness, and cutting depths all require speed adjustments to prevent damage to the bits, which are relatively expensive. Most mills are rigged with spray coolant systems that direct a fine mist of oil at the cut zone and the bit to prevent damage caused by overheating.

CNC mill cutters, many with collets attached to allow for rapid automated tool changing.

The tables used on both vertical and horizontal mills are usually around 10" × 50", although much larger and smaller examples are not uncommon. The table is usually grooved with a series of slots that permit a wide variety of clamping conditions as well as the draining of cutting fluid. Work can either be attached directly to

The bar is cut on the lathe using various tools.

Sections of aluminum round bar cut to size.

Coolant is automatically applied to the cut zone.

The finished parts are stacked.

the table using clamps or bolts, or fitted into a vise that is itself attached to the table. Mill tables are ground to be extremely flat and must be carefully mounted to ensure that the spindle and table are precisely parallel.

Aside from vises, two common attachments used to hold work pieces on mill tables are the dividing head and the rotary table. A dividing head is used to hold pieces that are incrementally turned and milled, like gear teeth. A rotary table is a swiveling table, much like a Lazy Susan, that makes it easier for a manual operator to mill curved slots.

Very large mills exist that have multiple spindles so that an operator needn't stop and reload to perform each new cutting task. There are also machines called traveling column mills, which have a turret mounted on a computer-controlled gantry. The gantry allows the turret to travel over very large stationary beds.

Cutters are made in a wide variety of shapes and sizes designed for specific cutting conditions. End mill profiles include square or rectangular grooves, concave and convex fillets, dovetail, and T-slots. Most end mills can be fed into material as drill bits are, and then directed at a right angle to form a slot; others can only engage the work piece at a right angle. Ball end mills have round tips to create rounded slots. Roughing end mills are cutters with serrated teeth for aggressive stock removal; they are usually used in conjunction with fine cutting bits for more precise work. Side cutters and slitting saws are shaped like wheels, with cutting teeth on their edges. They are mounted to arbors (shafts) and used for

A machined part. Note that the paths taken by the cutting tools to cut this part from a solid piece of metal are visible on the surface.

cutting narrow slots. Side cutters of various sizes can be ganged together to create complex profiles simultaneously. Fly cutters use a single cutting tool mounted to an adjustable arm that projects from a central shaft. By adjusting the arm, the cutting tool's distance from the shaft is altered, thereby increasing or decreasing the diameter of the cut. Fly cutters are used to bore circles or create round grooves (in which case the process is called trepanning), or to plane flat surfaces.

Different metals pose different challenges in milling. Aluminum is soft and easy to mill, but this softness can cause the metal to clump onto cutting tools, clogging their cutting edges or binding them and causing tool breakage. Most stainless alloys are hard and require slower cutting speeds and more coolant, although some alloys have sulfur added to them to make them softer (see Chapter 1).

Mills are typically used to produce parts with precise dimensions, especially those involving motion, such as gears and other mechanical components. They are also used to produce parts that must precisely interlock or that require accurate reveals that cannot be produced cleanly using any other process. Very often, milling operations are only performed on a specific section of a part that was made using cruder fabrication methods.

CHEMICAL MILLING AND ETCHING

Chemical "milling" is another way to produce intricate parts with precise dimensions using an entirely different process from the one described previously. In chemical milling, a template in the shape of the desired part is made from photoresist and applied to the surface of a metal plate. Photoresist is a resin that hardens and bonds to the metal plate when exposed to ultraviolet light. Once the template has been bonded, the plate is immersed in an acid bath, and the acid removes (etches) the uncoated metal, producing a part in the shape of the photoresist. There are also methods that use similar coating materials, called maskants, that do not need light exposure to harden. Chemical milling is particularly suited to the production of very thin, flat parts. Etching can also be performed to create embossed textures in the surface of a metal part.

THE LATHE

In some respects a lathe is the opposite of a mill. With a mill, a spinning cutter removes stock from a clamped work piece. A lathe causes the work itself to spin, and the operator removes stock by applying stationary cutting tools. In some ways, a lathe is to metal what the potter's wheel is to clay; the raw material is spun, and cutting tools are applied to generate shape. Although the metal part spun on a lathe is not plastic like clay, the exterior profiles of thrown pottery are very similar to those created on a lathe. Rather than

How a spinning part
is cut on a lathe.

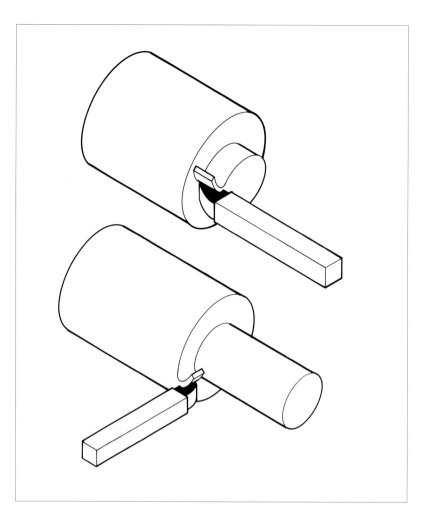

Lathe.

adjust the work piece position, the cutting tool itself is gradually moved back and forth and also laterally. Lathes can be used to create very complex shapes, but in most cases a cross section taken at any point will be circular. It is useful to think of parts made on a lathe as being composed of strings of discs that can vary as needed in diameter and thickness. This means that pieces can be notched, smoothly tapering, or even circular in profile. Lathes can also be used to cut threads, tap and drill holes, and even perform certain milling operations.

Lathes are categorized according to two dimensions: their swing and their length between centers. Swing refers to the maximum diameter of a work piece that can be placed on the lathe. The length between centers refers to the maximum allowable length of a work piece secured on the lathe.

Rotation on a lathe is transmitted from an electric motor via belts and gears to a circular protrusion called a spindle, which is mounted in the headstock. The spindle is what spins the work. Opposite the headstock and mounted on a grooved base called a bed is a piece called the tailstock. The tailstock holds a sharply pointed rod that is precisely aligned with the center of the spindle. This pin is called a "dead" center if it is stationary and a "live" center if it is mounted in bearings in order to spin. The work is securely set between the headstock and the tailstock. At the tailstock side, a small hole is drilled in the center of the work so that it will spin around (or with) the center pin. The tailstock can be moved closer or farther away from the headstock to accommodate the size of the work piece.

At the headstock, where the rotation is actually applied, there are several ways to attach the work. A faceplate can be attached to the end of the work and positioned securely into a clamping chuck attached to the spindle. The chuck is a large round plate with a set of three or four jaws that securely hold the spinning part. If the part is very short, it may be held only by the clamp, and the tailstock is not used. Very small parts can also be held on the headstock side with a screw-on chuck, similar to the chuck used on a drill press. Parts can also be attached to the headstock by removing the chuck entirely and screwing a faceplate directly to the spindle. A faceplate is a round plate with a series of oval holes through which clamping screws are placed. Sometimes a device called a lathe dog is used with a faceplate. A lathe dog is a circular ring with a threaded screw pointed into the ring and a bent "tail." The tail inserts into a slot in the faceplate, and the work is fitted through the ring and secured by tightening the screw.

The cutting tools used with a lathe are called bits. Lathe bits are usually carbide-tipped, single-point cutting tools that possess a variety of profiles to match their cutting purpose. Different tools are made for shaping the outsides of work pieces, for cutting internal and external threads, and for boring out holes.

Manual knurling tool.

Knurled finish on a tool, applied on a lathe.

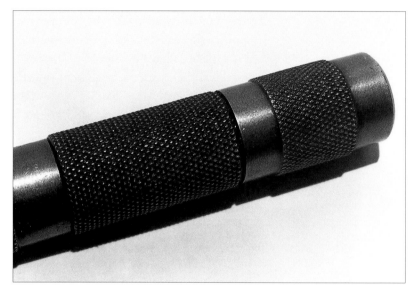

Bits can be fed into the work either manually, by turning controls, or "automatically" by allowing the lead screw to control the speed at which the bit travels parallel along the work. The lead screw is a threaded rod that runs the length of the lathe and is spun by gears from the headstock. The mobile platform on which the bit rests can engage the lead screw and translate its spinning into motion along the length of the lathe. There is a direct relationship between the rates at which the spindle and the lead screw are turning, so they can be calibrated to obtain a desired result. When cutting threads, it is critical that the bit move at an appropriate speed along the work piece to ensure that the threads are properly spaced. Imagine a spinning rod and a pen held at a right angle to the rod. If you ran the pen very quickly against the rod as it spun, you would find a very wide spiral drawn on the rod's surface. Similarly, if you moved the pen very slowly, the spiral would be tightly coiled along the rod. The same principle is at work when cutting threads on a lathe.

For most designers, the need to have custom threads cut on a lathe will probably not arise, since threaded rod is widely available in many lengths and materials. The same techniques used to create threads can be employed to make any sort of spiral profile that might be desired.

The lathe provides the designer with a vocabulary of fluid, tapering shapes that stand in stark contrast to the uniformity of standard extrusions. The profusion of urns, undulating finials, and balls decorating every protrusion on neoclassical architecture and furnishings was one approach to the product of the lathe. Perhaps eighteenth-century minds found the perfect radial symmetry of these shapes expressive of their vision of a universe spinning in eternal balance. Art Deco metalwork abounds with table bases, lamps, and finials cut on the lathe that are intended to express the beauty of machine precision itself, free from the imperfections of handwork. Computer control of lathes will undoubtedly lead to new vocabularies of form in our own time.

Additional CNC parts cut on a lathe, with interlocking threaded and tapped elements.

Parts cut on a lathe using CNC.

WELDING AND SOLDERING

Welding and soldering constitute one of two basic strategies employed to assemble metal parts into larger wholes. The second strategy, mechanical fastening, will be discussed in Chapter 9.

WELDING

Welding is the assembly of parts with the application of heat sufficient to cause melting at the joint between the parts. When resolidified, the melt zone creates a unified bond between the two parts. The heat necessary to melt metal is generated by several processes. These include oxygen/fuel welding, electric arc welding, resistance (spot) welding, friction welding, and laser and ultrasonic welding. Of this list, oxygen/fuel and electric arc welding are the most commonly used processes for smaller production runs.

OXYGEN/FUEL WELDING

Oxygen/fuel welding involves the mixing and burning of a controlled combination of oxygen and a fuel gas, usually acetylene. The gases are mixed in a torch and released in a concentrated stream through a small nozzle at its end. A handheld flint, which is struck with a small file in a cup, is used to ignite the gases. Regulators attached to the separate gas tanks (called cylinders) control the amount and flow rate of each gas released. Varying the mix and flow rate can substantially increase the temperature of the stream, and the same equipment used to weld pieces together can also be used to cut through solid pieces (see Chapter 3).

As the flame melts the metal the surface edges are rapidly oxidized, possibly interfering with welding and causing messy connections. To prevent oxidation buildup on metals and also to initially clean the joined surfaces prior to welding, compounds called fluxes are used. Flux is usually applied as a paste and can be made of tallow, rosin, zinc chloride, or other chemicals, depending on the materials to be joined. Usually a small V-shaped recess is ground away at the point of contact between parts to be welded. This provides a spot for the weld to puddle and form a bond with more surface area on both parts. Extra material, called filler, is often added to the melt zone

to replace metal removed from the recess. Filler comes formed into wire, and is usually made of the same material as the joined pieces.

Oxygen/fuel welding is fairly simple to learn, and uses relatively inexpensive equipment and consumables (those items that are used up by the process, such as gas, flux, filler rod, and torch tips). It is useful for joining softer, nonferrous metals, such as bronze and brass, but it has limited applications for steel (except in the case of cutting, where the process can be employed to cut plates as thick as two feet).

ELECTRIC ARC WELDING

Electric arc welding is by far the most common and versatile process used for welding metal, especially steel. The process can be used with stainless steel, aluminum, and titanium, but typically not with copper-based metals such as bronze and brass. As a general rule, only pieces made from the same material can be welded together to form a solid joint. Although there are many variations, the three most often used methods of electric arc welding are referred to in common parlance as MIG, TIG, and stick welding.

MIG (which stands for "metal inert gas") welding is initiated by sparking an electric arc between the welded parts and a consumable electrode. The unattached parts are then clamped tightly together. In a fabrication shop, the parts are placed on a flat, level layout table. The electric arc travels from the electrode in a handheld torch, through the welded pieces, and into a clamp (called a ground), which is either attached directly to the welded pieces or to the layout table (if the table is also metal). The consumable electrode is spooled metal wire, which is fed through the torch. The electrode is protected by a blast of shielding gas (often argon), which isolates the electrode and the weld zone from the air. Argon is ideally suited for welding because it will not dissolve into the molten metal. For this reason, it is considered "inert." As the arc melts the edges of the welded material it also melts the wire, which serves the same role as the filler rod in oxygen/fuel welding. The molten zone, called a puddle, is rhythmically advanced to create a series of overlapping layers that (ideally) create an even and consistent appearance. Without the shielding gas, the weld puddle would draw oxygen from the air, which could create oxidation and interfere with the formation of a strong weld. The torch gets very hot, so it is kept cool with a circulated stream of air or water.

This is a special jig used to rotate heavy parts during welding, so that the operator does not need to lift the part to work on all of its sides.

MIG welding is a fast, relatively predictable assembly process,
which, when performed by a skilled welder, can produce very strong
welds, especially on steel and stainless steel. MIG welds themselves
are somewhat large, uneven, and often have a coating of slag (glass)
deposits created by atmospheric impurities that circumvent the
shielding gas, enter the melt zone, and must be chipped away.

Far more attractive welds are produced by a process called TIG.
Like MIG, TIG (for "tungsten inert gas") welding uses an electric
arc generated between an electrode and the welded parts to create
a melt zone. Unlike MIG, the electrode is not the filler material.
Instead, the electrode is a stationary rod fixed at the end of the
torch, which is made of solid tungsten. The tungsten is not deposited
in the puddle, but must be ground periodically when its tip melts
or becomes contaminated with splattered metal. The filler material
is supplied in rods that the welder holds in his other hand and
touches to the arc to melt. As with MIG, an inert shielding gas such
as argon or helium is released from a pressurized cylinder through
the torch nozzle to shield the puddle. Steel filler rod is coated with
a jacketing of copper, which functions as flux.

TIG welding provides the operator with substantial control over
the appearance of the weld, but when compared to MIG the price
for this control is a reduction in speed. If clean, evenly lapped puddles
are your goal, TIG is the process to specify.

Stick welding is more precisely but infrequently called shielded
metal arc welding (or SMAW). With stick welding, a stationary,
consumable electrode, or "stick," made of filler material coated with
flux is used to conduct an arc to the weld zone. As the stick is

Stick welding with a small portable welder.

consumed, it throws a considerable amount of molten metal spray into the air, which is why welders often refer to the sticks as "sparklers." Stick welding does not require the use of shield gas, which makes it useful in remote locations where it might be difficult to bring a heavy gas tank. The lack of shielding gas results in large amounts of slag, which forms a thick coating over the entire weld and must be chipped off.

Stick welding is easy to spot at night on construction sites owing to the distinctive flashes thrown from the sparklers. The visual quality of stick welding tends to be poor when compared to TIG, but the structural quality is excellent. This makes it useful for unseen joints, or joints where structural strength is the only criteria.

SPOT WELDING

Spot welding, more specifically called resistance welding, is a process used to join thin sheets of metal together. The sheets to be connected are sandwiched between two tightly clamped electrodes. When current is run through the electrodes, the resistance to the arc generated by the material creates a localized melt in the sheets beneath the electrodes that causes the pieces to be joined. Spot welding is particularly well suited to automation, although all welding processes can certainly be automated for production runs.

FRICTION WELDING

Friction welding is a rather exotic process in which parts are combined by spinning one piece at a high speed and pressing it tightly against a stationary piece. The friction generated by their contact creates heat, and the parts are thereby joined. Another process, called laser welding, relies upon a laser to supply the heat necessary for melting. Ultrasonic welding is yet another process that depends on energy transmitted by sound vibrations. These welding processes are well beyond the reach of all but the largest industrial concerns at this time, but then again, at one time, so was TIG welding.

WELDING AND DESIGN

There are a number of critical issues to keep in mind when designing welded objects. First, welding both generates and releases stresses that can cause substantial warping, especially in thin materials. Extensive welding may require that thicker material be used. This is particularly true with stainless steel and aluminum. To counteract warping, the fabricator carefully clamps each work piece to hold it rigid while the welds cool and contract. The fabricator must exercise considerable expertise and judgment in determining how pieces are clamped together, the order in which joints are welded, and the amount of time welds are allowed to cool before releasing clamps.

A temporary jig made up of a number of guides and clamps that serve to position parts to be welded, and keep them from moving and warping.

A very good welder can often anticipate how welds will affect a piece and actually "overclamp" a piece slightly, so that it will spring into the correct position when it cools. It is often wise with welded objects (especially skeletal or "frame" constructions; see Chapter 15) to design a way that they can be clamped or fastened back into position at their final location. For example, a table base may rely on its connection to a rigid top to return to a level and true configuration. A large sign or sculpture may require some clamping when it is attached to a wall or a fixed base to assume its intended shape. These are not design or fabrication flaws; rather, these are the inevitable results of the welding process.

Another major consideration is the appearance of the weld itself. Depending on the aesthetic inclinations of the designer, an unground MIG or stick weld may seem either appropriate or too crude for a finished piece. Similarly, many architects routinely specify that "all welds shall be ground smooth," forgetting that a weld can be compromised if too much is ground away. A better solution might be to locate unground welds in concealed areas, or, possibly, to specify exposed TIG welds, evenly lapped, where they must be visible.

For structural welds there is a system of symbols that can be used in drawings to precisely express weld configurations, size, and depth of penetration. The specific criteria of structural welds are probably of more interest to engineers and fabricators than to designers, but it is important to understand what various weld conditions will look like. The most basic joint created between two plates laid side by side is called a groove weld. To be strong, a groove weld should be applied to both sides of a joint, and the plate edges should be beveled slightly prior to welding to increase the puddle zone. A weld like this one, fully exposed on a flat surface, is the easiest weld to grind smooth. Another easy weld to grind is called a plug, or "rosette" weld. Plug welds are created by overlapping two plates, drilling a hole through only one plate, and using the hole as a weld point. The hole should be overfilled with melted material. Plug welds can then be easily ground flush to the surface of the top plate to create a concealed joint. Welds at outside corners are also very easy to attack with grinding tools.

The most difficult welds to grind are called fillet welds. Fillet welds are placed at the inside corner of perpendicular plates, where only burs and abrasive stones can easily reach them (see Chapter 10). Another common location for a fillet weld is at the joint between a plate and a round tube or pipe perpendicular to it (think of the typical steel table leg and mounting plate). Grinding fillet welds smooth is time-consuming, so they are frequently left unground. On painted metal objects, it is not uncommon to discover that fillet welds have been smoothed artificially with auto body paste. If two assembled pieces of metal are mitered (cut at a 45-degree angle), it is often possible to leave the inside corner unwelded and rely on welds at the exposed sides to hold the pieces together.

A groove weld between two flat plates. To be strong, the weld should be made on both sides of the plate.

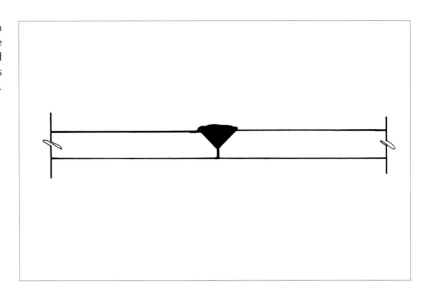

A rosette or plug weld. A very handy weld for decorative work, since it can be ground smooth for concealment.

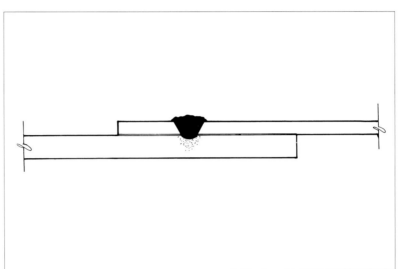

A fillet weld. A fillet weld should also be made on both the inside and the outside of the joint.

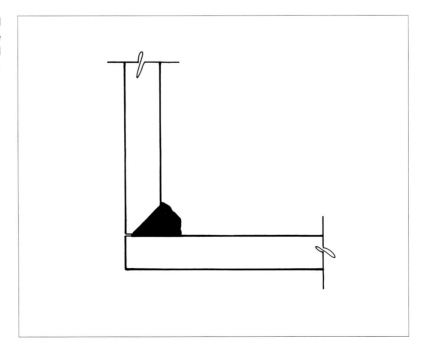

WELDING SAFETY

With all electric arc welding, two particularly dangerous by-products are produced: poisonous metal fumes and retina-fusing light. While the effects of the fumes may be gradual or cumulative over time, there is nothing gradual about the blindness that can easily be caused by looking at an unscreened arc for even a few seconds. It is far from the purpose of this book to provide even a cursory list of shop safety rules, but we implore you to remember that you must never look at any welding arc without a mask or screen specifically made to shield you from the light. A welder typically wears a leather apron and long gloves to protect from molten metal splatter, along with a welding mask fitted with a powerful light filter. Fumes are evacuated by localized extractors with hoods that can be moved closely to the melt zone.

SOLDERING

Soldering and brazing are the processes of joining metal parts with a melted filler material. Unlike welding, only the filler is melted; the joined parts remain solid. The filler holds the parts together by adhesion only, so the joint is definitely non-structural. The difference between brazing and soldering is rather subtle; solder filler melts below 800 degrees Fahrenheit, and brazing filler melts above 800 degrees Fahrenheit.

Solder can be purchased in bars, slabs, and ribbons, but the most common form is wire. Most solder is tin- or lead-based; in some cases it is made with a flux core made of rosin or tallow. Solder is melted with the use of an iron, torch, or electrically heated metal stylus, much like a calligraphic pen. Almost any metal can be soldered, but it makes the most sense to use it to join metals that are either inherently difficult to weld (copper and its alloys), or too small or too thin to survive even partial melting. Soldering is also useful for joining dissimilar metals that cannot be welded together (copper to stainless steel, or brass to aluminum, for example). Soldering is a handy method for creating watertight joints, and it is frequently used in plumbing and roofing. Since the solder is liquefied, it can be made to fill very small joints since it is subject to capillary action.

Brazing fillers are made from materials with higher melting points than lead and tin, but still lower than the metals they are used to join. Examples include phosphorus/copper, zinc/copper, aluminum/silicon, and silver. Fluxes are used to clean the attached surfaces of oil and grease, to prevent oxidation, and to limit the generation of fumes. Because of the higher melting point of these compounds, the oxygen/fuel torch is a good heat source for brazing. Handheld propane tanks, equipped with torches, can also be used.

MECHANICAL FASTENING

Mechanical fastening is the act of joining parts with the aid of devices such as screws, rivets, and pins. Fasteners are inserted into overlapping holes in separate parts to create a compressive or frictional bond. Most fasteners are designed to be removable, thus an assembly held together with fasteners can be taken apart without damage to the individual components. This is a significant advantage over a welded connection in situations where specific parts might require replacement or removal, as in machinery or building assemblies.

The most basic mechanical fasteners are externally threaded screws and bolts. A screw or bolt is one part of a larger system that includes a nut and a pair of washers. The shaft of the fastener is cut with spiraling grooves called threads. The nut, which receives the fastener, has a set of inwardly cut threads that are closely mated to the outwardly directed threads of the screw. As the nut or the head of the fastener is turned, the threads engage and cause the nut and

Fasteners, nuts, and washers.

the top (head) of the fastener to come together. Screw heads, bolt heads, and nuts are designed to fit a wide variety of turning tools such as wrenches and screwdrivers.

Notwithstanding the previous description, the boundary between fasteners defined as screws and fasteners defined as bolts is somewhat unclear. Generally, bolts are meant to be fastened by turning the nut and holding the bolt head steady. Screws are usually the reverse; the head is turned and the nut is held steady. Also, bolts are only intended for use with nuts, while screws can be inserted into threaded holes in the attached material.

When the threaded fastener and the nut are tightened, they create a clamping condition that holds the pieces between them together. The nut and fastener cannot be detached by simply pulling them apart; they must be twisted, and even then, the twisting must be in the correct direction (usually counterclockwise). Possible ways for the connection to fail include the screw or bolt breaking, the

Screw heads are made for a variety of drivers. Top row, left to right: flat, Phillips, hexagonal, square (Robertson), torx. Bottom row, left to right: tamper-proof flat, spanner, Phillips with tamper-proof pin, tri wing, and notched head.

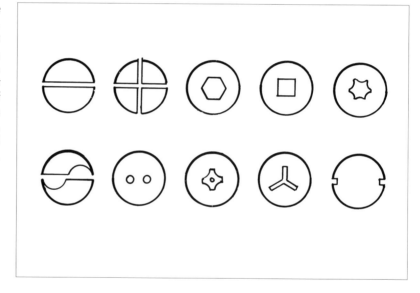

Various types of screw heads. From left: flat, cap (round or hex), button, and fillister. This group is far from exhaustive.

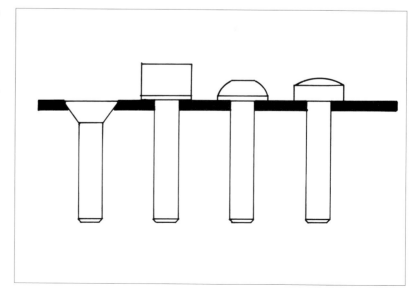

interleaving threads shearing from the main bodies of the screw and nut, the clamped material tearing, or vibrations causing the fastener to loosen. This last situation is very common with machinery. In order to keep the fastener from pulling through the material, the head of the bolt or screw and the body of the nut are larger than the holes they align, creating seats that distribute the force of compression. To add more force distribution, and frequently for cosmetic purposes (in the case of rough or misaligned holes), a pair of washers are usually added: one beneath the head of the fastener and another between the nut and the clamped material.

<table>
<tr><td>THREADS</td><td></td></tr>
</table>

THREADS

Threads on fasteners are sized according to the diameter of the blank rod into which they are cut and the number of threads per inch (or centimeter). They can also be described by their class of fit, a measure of how tightly or deeply the male thread of a particular size engages its female counterpart.

The most common thread sizing system for fasteners in the United States is called the American National Standard for United Screw Threads. The standard utilizes a 60-degree rounded V-shaped thread. Thread sizes are categorized by two numbers: the first is the diameter of the blank rod into which the male thread is cut, and the second is the number of threads per inch. The smallest diameters are described by code numbers ranging from 0000 (.021" diameter) up to 12 (.216" diameter). After 12, the next sizes are classified by their actual diameter, which in this case starts at $1/4$". This size is followed by $5/16$", $3/8$", $7/16$", $1/2$", $5/8$", $3/4$", and 1". Larger diameter fasteners certainly exist, but they are not stocked as commonly as the sizes listed. For the number of threads per inch, most screw manufacturers provide two types: a coarse thread and a fine thread. For example, size 6 screws can most often be found with either 32 (coarse) or 40 (fine) threads per inch, referred to as 6-32s or 6-40s. A third number, the shaft length of the fastener, is also specified when ordering. Thus, one may speak of a $1/4$"-20 × $5/8$" (a quarter twenty by five eighths—quarter-inch diameter, twenty threads per inch, five-eighths of an inch long). Virtually all screws and bolts are measured from under their heads to their threaded tip, with the exception of flat heads, which are measured from the top of their head to the tip. The most common head configurations are flat, cap (round or hexagonal), button, and fillister. The decision of when to use fine or coarse threads is an issue that arises in only the most critical engineering applications.

Be careful not to confuse the thread sizes used for the ends of pipe with the system used for fasteners, as they are entirely unrelated. Pipe threads are sized according to the outside diameter of various pipe, which you may recall is itself sized by inside, not outside diameter (see Chapter 2). Also note that there are sharply threaded screws, called self-tapping screws, that cut their own thread into thin steel sheet. Their threads are not calibrated with the same kind of accuracy as other screws since they do not need to match any sort of pre-tapped hole or nut.

The male threads of screws are created by either putting a solid, round rod on a lathe (see Chapter 7) and cutting the threads with a tool applied to the spinning rod, or by inserting a stationary clamped rod into a thread-cutting die which is then turned against the rod. The die is a circular cutting tool with a set of increasingly sharp teeth that carve the spiral thread pattern into the rod. The die is secured in a two-handled wrench. In practice, dies are probably more frequently used to repair dulled or damaged threads than to cut new ones. Better quality dies are split at one side and secured with a screw, both to relieve stress and to allow slight adjustments of the depth of the thread cut by tightening or loosening the screw.

The female threads of a nut are usually created with a tap, an implement about the size of a drill bit with a series of increasingly sharp cutting teeth. Internal threads can also be cut with a stationary tool inserted into a spinning part on a lathe. The far more typical process, however, involves inserting a spinning tap into a clamped piece. Most taps are not self-drilling; they are meant to be inserted into a pre-drilled, punched, or milled hole slightly smaller than the diameter of the tap. The tap then cuts the thread into the excess material as it passes (gently) through the hole. Taps can be used with hand tools called tap wrenches, which require the user to gingerly twist the wrench as the tap cuts the thread. Better quality tap wrenches are ratcheted for faster turning and have sliding handles

Taps and dies. From left: die wrenches, circular dies, a tap wrench and tap, and a variety of taps.

An automatic tapping machine attached to a drill press.

for awkward setups. Even with great care, experienced users of tap wrenches will break smaller taps with discouraging frequency. For production runs, there are automatic tapping machines that can be mounted to drill presses or mills. Automatic tapping machines are geared to adjust the turning speed of the tap for optimal cutting, and have clutches that can be set to disengage or spin in reverse when the tap encounters too much resistance, a feature that dramatically reduces tap breakage.

A nut is not the only place a female thread can be placed. A threaded hole can be situated in one of the two parts meant to be joined, eliminating the need for a nut. The only prerequisite for this detail is that the tapped material be sufficiently thick that enough threads can be cut to produce a strong joint. Threaded holes are particularly useful when one side of an assembly is inaccessible.

Screws and bolts are categorized by their specific use, the type
of tool used to tighten them, their head shape, and their strength.
Most fasteners are designed to satisfy the basic scenario already
described, with set screws being the one significant exception. Set
screws are usually headless and are meant to fit into a tapped hole
in one material and hold the adjacent material by friction applied
at their tip. They are normally tightened with a hexagonal driver,
which fits into a milled hole at one end of the set screw shaft. For
additional holding power, the tip is sometimes screwed into a hole
or depression in the held piece. Special set screws are fitted with
nylon tips so they can be fixed against fragile materials like glass.
Set screws are frequently used to secure adjustable shafts.

A section showing a
set screw holding
a shaft in place.

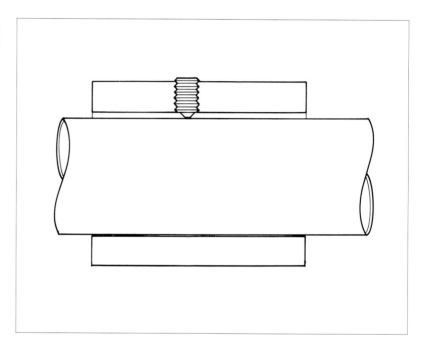

BOLTS Bolts are made from heat-treated carbon steel and stainless steel.
Typically, they have hexagonal or square heads, shapes that are
easy to hold with a wrench. They are intended for structural use,
so they are carefully graded by strength. Some bolts have round
heads with teeth on the underside, which allow them to grip
against the clamped material while their nut is tightened. T-slot
bolts have rectangular heads made for fitting into milled slots
on machinery.

MACHINE
AND CAP
SCREWS

The most common screws available are called machine screws.
They are usually fully threaded and can be turned with a variety of
screwdrivers, including the common flat, Phillips, hexagonal, and
torx styles. Of the four basic head styles, button and flat are most
common. Machine screws are intended for relatively low-strength

applications. They are frequently made from soft metals, such as brass and non-heat-treated steels. If the screws are made from steel, they are plated with a protective metal coating, such as zinc or cadmium. For situations where the strength of the screw is critical, there is a category of fasteners called cap screws. Cap screws are carefully rated for their tensile strength, which is typically brought to levels between 90,000 and 120,000 pounds per square inch (p.s.i.) by heat treating. Like many bolts, cap screws have hexagonal heads. Unlike bolts, the cap screw head is meant to be turned with a wrench rather than held tight. Torque wrenches, which are ratcheting wrenches with controls to measure and limit the torque they apply, are used for critical applications in machinery.

Socket head cap screws have circular heads that are turned by inserting hexagonal drivers into a matching hole in the center of their head. They are among the most elegant and well-machined of all fasteners, and typically the most expensive. They are made with cap, button, and flat heads, in heat-treated steel with black oxide, stainless steel, or titanium finishes. Socket head cap screws are rated for extremely high p.s.i. strength, in some cases as high as 170,000. Like all cap screws, longer screws tend to be threaded only partially along the length of their shaft. A common rule of thumb used by screw manufacturers is to make the length of threading equal to twice the screw diameter plus $1/2$".

SECURITY SCREWS

In applications where security is an issue, there are many kinds of tamper-resistant screws. The strategy employed by most of these is to require an exotic screwdriver to turn the head. Examples include "one way" screws, which can be screwed *in* with common tools but not *out*; spanner heads (two small holes engaged by a pronged driver); tri-wings, which are turned by a driver with three rotating blades; and even pin-protected shapes that require a corresponding hole in the driver to turn them. For obvious reasons, new styles appear with frequency.

Ordinary screws and bolts can be fastened more securely with the use of thread-locking adhesives or with nylon patches added at the factory to bind screws to their threaded hole or nut. For applications where vibration is extreme and a loose screw could cause catastrophic damage (such as an airplane), screws are made with tiny holes drilled through their heads so that they can be retained by a wire and easily located if they are shaken free.

NUTS

The majority of nuts are hexagonal so that they can be turned or held with wrenches. They are either open-ended or close-ended, in which case they are called acorn nuts. Acorn nuts are used either to

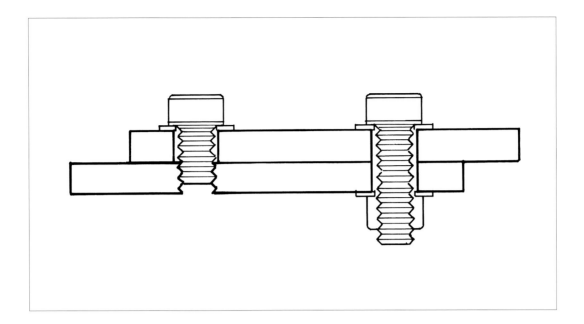

Two ways to use screws to assemble plates. On the left, the bottom plate is tapped. On the right, the fastener passes through holes in both plates an is held with a nut. Note position of washers between fastener and plates.

cap the end of a thread and keep it safe from damage, or simply because they create a more attractive joint than open nuts. Acorn nuts conceal variations in the amount of thread exposed and enforce a certain uniformity of visual pattern in assemblies. It is important to remember, however, that an acorn nut can only conceal a finite amount of the screw, so more careful planning is required than might be when using open nuts. Acorn nuts can be found with both low and high dome heads.

Another important category of nuts are those made to be loosened or tightened with fingers only. These are divided into wing nuts and thumb nuts. Wing nuts can be found in open or closed varieties and have a pair of ear-like protrusions cast into their sides. Thumb nuts are round and have a knurled edge so that they can be turned like a knob.

Panel nuts are round with a thin hexagonal flange that allows them to be inserted into holes for a flush fit. Lock nuts are made with nylon inserts or integral washers that clinch into the fastened surface, or with tapering threads that bind the inserted screw. Slot nuts are square or rectangular, and, like slot screws, can be inserted securely into milled channels.

WASHERS

Washers are usually round rings made from a wide variety of materials. Washers have four basic purposes. First, they spread the forces generated in a tightened screw over a larger area than the head of the nut, helping to prevent punching shear. Second, they conceal rough or misaligned holes. Third, washers isolate dissimilar materials to prevent galvanic action. Fourth, washers can aid in locking fasteners into position and resisting vibration. Locking washers include serrated (or toothed) washers that cut into the fastened material when clamped, and lock washers, which are made from wavy or split springs. As the spring is tightened, it stores torque to keep the screw fixed.

STUD WELDING

An interesting variation on screw fastening is made possible by a device called a stud welder. A stud welder uses electrical discharge to weld individual studs (small, headless, threaded rods) onto the surface of steel, stainless steel, or aluminum parts. The stud is placed in the barrel of a discharge gun, leveled, and spot welded into position. The stud can then be matched to a hole in another part and assembled with a nut. Stud welding is very useful for applications where one side of an assembly cannot be accessed (see Blind Fastening below).

Stud welding gun in action.

The availability of relatively inexpensive, highly reliable screws, bolts, and nuts has eclipsed the use of another major category of metal fasteners known as rivets. A rivet is a headed pin that is inserted into mated holes, and then crushed, so that the tip of the rivet is mushroomed to create a second head. Early rivets were heated cherry red, inserted, and then hammered by hand to create a permanent fastening. An amazing example of this kind of assembly can be seen at the San Francisco Maritime Museum, where the clipper ship Balclutha, built in 1886 and 302 feet long, was assembled from steel parts almost entirely in this manner. Today, most of these connections would be made with screws or bolts if disassembly for maintenance was desirable, or welded for permanent joining.

Rivets live on for one specific application where screws may not always work: blind fastening of thin sheet or untappable materials. Blind fastening means that only one side of an assembly can be

Rivet.

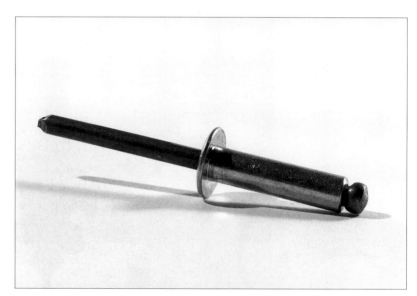

A rivet is inserted into a hole (left), and then the mandrel is pulled upward by a rivet gun to create a fastening.

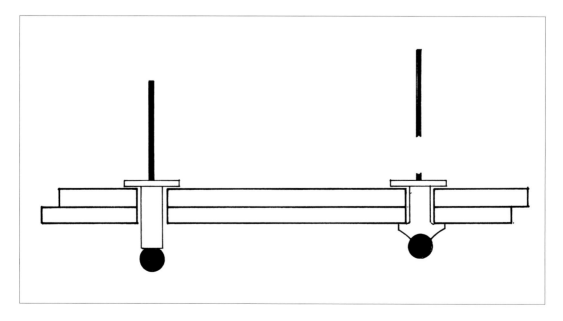

accessed. A typical rivet used today consists of a long, nail-like tail with a ball tip. The tail is called a mandrel. The mandrel passes through a tube with a washer-like flange, called the rivet body. The entire assembly is placed ball end first into matched holes, and force is applied by a rivet gun to pull the mandrel upward. The ball at the end of the mandrel catches the bottom of the rivet body and crushes it, creating a flange that cannot be pulled back through the hole. The mandrel then snaps off, leaving the ball trapped in the rivet body.

PINS   One other category of fasteners remains, and that is pins. Pins are metal dowels without threads that are fitted into mated holes. Pins can be tapered or made from coiled springs, so that they transfer forces at right angles to their primary axis to hold tight as they are hammered or hydraulically placed into position. Cotter pins are made from a pair of prongs that squeeze tightly against their mounting hole. Quick-release pins are made with retractable ball bearings at their tips, sometimes controlled by a spring-loaded button or handle at the exposed end, which is pushed to remove them from a hole.

The classification of fasteners is so vast and interrelated that it begins to take on the characteristics of evolutionary biology rather than a man-made system of tools. There are bizarre hybrids and periodic exemptions to almost every rule discussed in this chapter. Having had the good fortune of working in Southern California, where the aerospace industry has left a vast legacy of industrial processes and spare parts, one of our favorite field trips is a visit to a military surplus hardware store, where unusual and even unexplainable hardware can be found relegated to dusty bins, awaiting some new mission.

BLENDING AND GRINDING

Blending and grinding are processes through which stock from the surface of metal is removed with the use of patterned tools and abrasives. The purpose of grinding is to take unwanted material away from the work piece. Examples include flattening deposited welds or creating fillets for welds, cutting contours into stock, and breaking sharp edges. Blending is a more delicate process, accomplished with the same tools and performed to create visually smooth transitions from ground to unground areas.

PATTERNED TOOLS

Patterned files of various sizes and shapes.

Patterned tools are usually made from heat-treated steel and scored with sharp crisscrossed teeth. They are sometimes coated with very hard materials to increase their strength. A file is a handheld patterned tool, usually made from heat-treated steel, which is scraped back and forth against work to remove stock. Files are made

with flat or curved faces for contouring, and with round, square, rectangular, and even triangular cross sections for cutting inside edges. They are graded by coarseness; a "bastard" file is for very heavy, rough stock removal, a "second" file is for intermediate work, and a "smooth" file is for fine finishing and blending. Files are usually meant for use on small work zones, and generally on edges rather than faces.

Another category of patterned tools are called burs. Burs are small, radially symmetrical bits faced with scored teeth attached to shafts. The shafts are inserted into collets on tools that spin the burs at very high speeds. The main tool used for spinning burs is called a die grinder. Die grinders are either air- or electric-powered handheld tools that spin the burs at anywhere from 12,000 to 80,000 revolutions per minute (r.p.m.s) depending on the model. Die grinders are light and can be positioned very precisely by a skilled worker.

Rough grinding.

Grinding tools. Clockwise from top left: air-powered die grinder and right-angle die grinder (with a selection of grinding stones beneath them); electric grinder/sander for use with 7" abrasive discs; a selection of paper-backed abrasive discs and rubber backing pads used to hold them; a 5" electric grinder; an air-powered belt sander; a loose belt; and a selection of small abrasive discs.

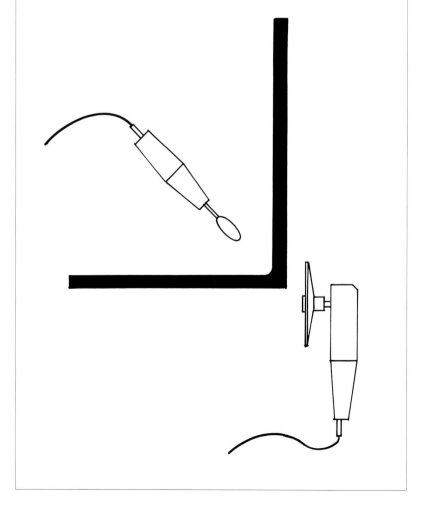

Different grinders are appropriate for different conditions. For outside corners, a right angle grinder can be used. For inside corners, stones or burs mounted to a die grinder are more appropriate.

The burs themselves are made in a wide variety of shapes, such as cones, ovals, and balls, so the user can match the shape of the tool to the contours of the grind area.

ABRASIVES While patterned tools are extremely useful, the majority of blending and grinding is performed using abrasives. Abrasives are hard (usually crystalline) materials capable of grinding metal. Not surprisingly, diamonds constitute the the best abrasive available. Tiny diamond crystals are placed in an epoxy suspension that is coated on the tips of cutting and grinding tools, where they perform excellently on the hardest of metals. Unfortunately, the less utilitarian market created for diamonds as jewelry makes this abrasive too expensive for routine grinding applications.

There are four crystalline abrasives commonly used in metalwork: aluminum oxide (the one most frequently used); silicon carbide; ceramic aluminum oxide; and zirconia alumina. All are microscopically sharp and capable of scoring a metal surface. Their primary drawback is that they are very short lived; as the crystals are heated by the act of cutting, they shatter, dull, or simply fall away. Some abrasives, such as zirconia alumina, have added grinding power because they shatter into smaller, but equally sharp bits that have continued cutting action.

Abrasive materials are crushed into small pieces, mixed with resin or glue, and cast into wheels and shapes (called stones) or attached to paper backings (primarily belts and discs). An extremely useful variety of abrasives is made by attaching the crystal/resin mix to webs of nylon fibers. Called non-woven abrasives, this product appears

A non-woven abrasive pad applied to copper.

in the domestic world as those handy green pot-scrubbing pads with the sponge attached. Non-woven abrasives are manufactured into hand pads, discs, and belts, and are very useful for blending and cleaning metal surfaces (see Chapter 12).

As stock is removed by an abrasive product, melted bits of metal (especially with softer metals such as aluminum) clog the crystals and impede cutting. The dilemma of clogging, which has confounded metalworkers for centuries, is partially (and only partially) alleviated by using tools called dressers. Dressers are handled tools with sets of star-shaped wheels at their end. When applied to the surface of a spinning abrasive wheel or belt, dressers will pull away some of the metal and expose fresh crystalline grains. Dressers with diamond-studded faces can be used to shape abrasive wheels and stones for special contours.

The coarseness of crystalline abrasives, the "grit," is carefully measured and abrasive products are sorted accordingly. The lower the number, the coarser the grit, and thus the more aggressive the stock removal. Grit measurements for aluminum oxide products usually start at 16, 24, 36, then 40, 50, and then jump to 60, 80, 100, 120, 150, 180, 240, 300, 400, and even 600. A product with a grit of 60 or even 80 is very coarse and suited for taking large welds down or grinding contours into the edges of a thick plate. A 120 or 150 is smoother and works well for blending on aluminum and steel. A steel part that requires grinding and will be painted should be taken to at least 120 to remove visible scratches that will affect the final coating. A 300 is a very smooth abrasive, only useful as a step in the process of polishing a part (see Chapter 12).

Grinding stones are abrasives manufactured in a wide variety of shapes that mirror the shapes used for carbide burs. Since abrasives are shed as they are used up, stones tend to begin as larger sized parts than burs. Some fabricators make their own custom stones for unusual profiles by grinding away part of a standard shape with a dressing tool. Abrasive grinding wheels and cups are attached to bench grinders. A bench grinder is a very simple machine; it consists of a motor, a shaft, and a base. Many bench grinders also have small pads for resting the work while it is ground against the edge of the wheel.

Abrasive paper-backed discs are manufactured in diameters from $3/4"$ to 20". Small discs (under 3") are used with a right-angle die grinder (a die grinder, which is made primarily for burs, has a collet at the end of the tool in line with its main axis, while a right-angle die grinder has a collet set at a ninety-degree angle). The right angle makes the application of the spinning disc to metal surfaces more natural to the motions of the wrist and provides the operator with greater control.

Medium-sized discs (4" to 5") are used with handheld electric grinders and sanders. The discs are usually attached at the center by an arbor, which screws through a hole in the disc onto a threaded shaft. A handy version of this size of sander is called a random-orbital

Grinding a continuous weld from a volumetric solid.

A random orbital sander on stainless steel.

sander. The random-orbital sander is used with adhesive-backed discs, which are attached to a pad that oscillates slightly off axis. The oscillation results in a non-directional, random swirl pattern that conceals scratches. Unlike regular handheld sanders, the random-orbital sander is almost foolproof for blending. Blending with electric sanders takes practice and patience, since the swirl pattern it makes can cut deeply and leave a distracting pattern if the tool is used aggressively.

Larger discs (10" to 20" diameters) usually have adhesive backing and are attached to large stationary tools called disc sanders. Disc sanders have a flat metal table mounted in front of the disc so that larger pieces can be rested at a controlled angle while being ground. Disc sanders are excellent tools for shaping since they remove large quantities of stock quickly.

Belts are produced for use with the same hierarchy of tools as the discs; small ones are used on air tools, medium ones on electric, handheld units, and the largest ones on stationary bench- or floor-mounted machines. The larger belt machines are useful for sharpening, shaping, deburring, and beveling, while smaller belts are used much like files to remove stock from tight intersections.

A 20" disc sander (left) and a large belt sander.

A small portable air compressor, used to run air tools.

Grinding is in many respects the source of some of the worst long-term health hazards caused by metalworking. Careful attention should be paid to protecting workers from the clouds of dust created by stock removal. As a designer, it is wise to try and minimize the amount of grinding necessary on a piece, since it is time-consuming and adds substantial cost. It is particularly difficult to grind welds at inside corners, since this often must be accomplished with burs and stones, which remove material much more slowly than grinding discs. Strategies that might be employed to keep grinding to a minimum include concealing weld points in unseen recesses or joints, or even simply letting clean, evenly lapped welds remain unground.

SANDBLASTING AND DEBURRING

Sandblasting and deburring are processes that use pellets of hard material to condition the surface of metal parts. The conditioning can be performed as a preparation for other finish work, as a way of removing sharp edges and burrs, or as a final finish in itself.

BLASTING

Sandblasting involves the use of silica sand or other fine particles to abrade unwanted paint, oils, oxidation, and light scratches from the surface of a metal part. Sandblasting is an excellent surface preparation for painting, since it both degreases the surface and creates a gritty texture to which paint can easily bind. On metals such as stainless steel and brass, sandblasting is also used as a decorative finish. A sandblasting machine consists of a hopper filled with sand, a vacuum hose, and a gun. The sand sucked up by the hose is mixed with compressed air and propelled through a tiny nozzle at high velocity. Because sandblasting is extremely dusty and messy, it is usually performed in controlled environments. The smallest space used for blasting is called a sandblasting cabinet, a box with a small viewing port and a latching door for placing and removing parts. The gun is positioned inside the box and the operator wears a pair of reinforced neoprene or leather gloves, which he uses to maneuver parts through a pair of small ports. The sand collects in the base of the cabinet, where it is gathered by the vacuum hose for reuse. Larger parts are sandblasted in enclosed rooms with airtight doors. In this case the operator wears a sort of "space suit" to protect from the spray and provide fresh air. Sand is collected and recirculated many times before being discarded.

The results of sandblasting are an attractive, matte, powdery finish on the surface of parts. Parts that have been blasted must be handled with care, since they are extremely susceptible to oxidation. Sandblasted steel will rust almost immediately if touched. On other metals, such as stainless steel, the surface will not rust but will collect dirt and oil that is very difficult to remove. Also, it is important to note that blasting will cause parts made from thin sheet (usually 16-gauge or less for steel) to curl and warp.

A very large room used for sandblasting.

Inside the sandblasting room.

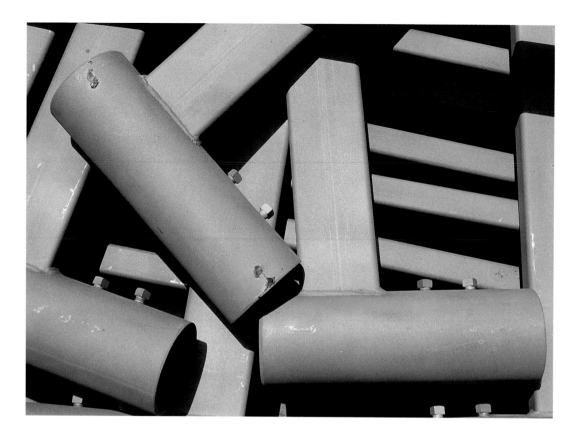

Sandblasted steel parts.

Other blasting media can be used, including ground nut shell, glass bead, and metal shot. Glass bead creates a smooth, almost polished finish. Metal shot blasting, also called shot peening, creates a subtle pattern of random dimpling. Shot is graded by size, and different gauges of shot create different finishes. Shot peening actually has a structural purpose; the tiny deformations created by the impact of thousands of shot balls dramatically increase the strength of metal parts designed for exposure to bending or twisting stress.

DEBURRING

Deburring is the removal of sharp edges and roughness from metal parts by immersing them in a vat of stone, ceramic, or plastic pellets that are either tumbled or vibrated with the parts. The movement causes the rocks to gently pound against the parts many millions of times, removing loose burrs and chips left from previous fabrication processes. Deburring can also be used to create smooth surface finishes ranging from a mottled speckle to a bright, reflective polish. The mottled finish is so attractive that it is frequently used as a final surface, especially on aluminum parts.

The shape of the vibrating or tumbling media is very important to a successful operation. There are numerous stock shapes of pellets, each designed so that they will strike as many surfaces of a particular piece as possible without becoming stuck in holes or tight corners. The process is very cost-effective, since thousands of parts can be finished at one time with minimal labor. Deburring

An aluminum table
with a deburred finish.

Deburring and ball
burnishing media
with typical parts.

usually takes several hours per batch, although the rate (and therefore the cost) of the process depends on many variables.

Parts are vibrated in large open vats, and tumbled in closed barrels filled with mildly abrasive liquids, small pellets, or just the parts themselves. When small steel balls are used as media in a tumbler, the process is called ball burnishing. Ball burnishing can be used to create very smooth finishes on small parts that would be difficult to polish by hand. Every vendor specializing in deburring will offer a slightly different selection of finishes and processes. For large runs, custom media shapes can be produced. The process can be applied to many materials other than metal, including plastics, glass, and especially stone.

Deburring vat.

Detail of a part in a deburring vat.

BRUSHING AND POLISHING

Brushing and polishing are methods for producing fine, uniform finishes on the surface of metal parts. Brushing, which leaves a finish consisting of fine parallel scratches (called grain), is often a first step along the way to polishing. Polishing creates a reflective (specular) surface, either with or without grain, which is produced using increasingly fine abrasives. Brushing or polishing can be done to parts in preparation for plating or anodizing (see Chapter 14), and sometimes the finishes stand alone on untreated parts. Parts made from steel generally must be plated after polishing, while stainless steel and such non-ferrous metals as aluminum and the copper alloys can in many cases be left uncoated, requiring only occasional maintenance to remove discoloration caused by oxidation.

BRUSHING

Brushed finishes can be created using a variety of processes. Grain can be produced on very rough stock by using coarse grit abrasive belts attached to belt sanders. While small or curved parts can be brushed using floor-mounted belt sanders, large, flat parts can easily be brushed using handheld belt sanders. This is accomplished by attaching the part to a flat surface with clamps (or double-stick tape if the entire surface must be brushed) and running the sander over the surface. It is a good idea to use a guide to keep the belt sander straight, since very rough material (like hot-rolled steel) will require several passes. Brushing is usually performed with a 60- to 80-grit belt. The brushed finish is slightly rough to the touch, and while a brushed part may have a sheen, it is not particularly reflective. If you are starting with a very smooth surface, a brush can even be applied by hand using sandpaper or non-woven abrasive pads mounted to a sanding block. A brushed finish can also be applied using a wire wheel spun on a motor.

Runs of many flat parts can also be sent through a large, specialized belt sander, called a Timesaver. A Timesaver uses very large abrasive belts (24" wide is typical) that are turned slowly under increasing pressure against pieces moving beneath the belt on a neoprene

Brushed flat bar.

A timesaver brushing
a group of parts
simultaneously.

conveyor. The brushed finish can be applied in one or two passes. Parts that have curved faces (like sinks) can be brushed very quickly using handheld grinders equipped with belts mounted over soft rubber "bladders" than conform to the curvature of a part.

POLISHING

Polishing is primarily decorative; the luster and reflection of polishing invites touching in a way that no other finish on metal can. It is eminently appropriate for anything that must be handled, such as door knobs, pulls, or utensils, but not suitable for parts that must actually be *held* as a matter of safety. Grab rails and steps, for example, should be textured to provide friction and improve grip. Polishing is also functionally useful, since smooth surfaces are easier to clean and less likely to hold moisture, giving them a measure of resistance to oxidation that rougher surfaces do not possess.

Polishing is produced by brushing a part with increasingly fine abrasives. For the first few steps on harder metals, abrasive discs and belts can be used to remove very rough spots. As the grit grows finer (usually above 320-grit), the abrasives are applied in paste or liquid suspensions, which are rubbed against the part using large cotton wheels called buffs. Buffs are spun on simple motors, which sometimes provide speed control for various effects. While increasingly fine amounts of material are removed by polishing, buffing does not cause appreciable stock removal; instead, a liquid flowing of the surface metal is created, producing a smooth luster on the outermost surface of a part.

Polished parts.

Polished parts.

Polishing a part using a cotton buff.

Polishing is an art in which there are multiple variables at play when performing the process. These include the type and size of abrasive, the speed and type of buff used, the orientation of the part to the buff, and even the sequence of abrasives used. The greatest variable, however, is the skill and attention of the craftsman applying the finish to the part.

Polishing can also be produced using several more technology-intensive processes. Laser polishing uses a brief pulse of laser light to melt the surface of a part. By carefully controlling the duration of the pulse, the surface is heated in such a way that the melt zone resolidifies into a smooth layer, much like remelted ice on a road. Electropolishing is another common process used, especially for large runs of smaller parts. In a process that is basically the reverse of electroplating (see Chapter 14), a part is immersed in a hot alkaline bath (an electrolyte) that is negatively charged. The part to be polished is positively charged. The interaction between the part and the bath creates a smooth finish on its surface. Polished finishes can also be produced on parts using vibratory deburring, ball burnishing, or tumbling, as described in Chapter 11.

SPECIFICATIONS A thorough specification system for polished (and other) metal finishes is produced by the National Association of Architectural Metal Manufacturers. This coding system for finishes is commonly used in the industry to communicate intended finishes, particularly with architectural projects. Since brushing and polishing are frequently the final processes used to finish stainless steel parts, some familiarity with the finish standards for stainless is very useful for designers, and so it is described in detail below.

For mechanical finishes on stainless steel (finishes produced using abrasives in polishing and brushing operations), the N.A.A.M.M. system describes a continuum, spanning from untreated hot-rolled surfaces all the way to highly polished finishes with no visible grain. The designation for a hot-rolled surface, which is dull and fairly rough, is #1. The next is #2, split into 2d and 2b, the "d" standing for a dull finish produced by cold rolling, and "b" for a bright, smooth, almost reflective surface created by running a 2d sheet through polished rollers. Although the 2b is somewhat reflective, the sheet is still perceived as gray-colored. The first level of finish considered "polished" is #3. It is produced with 100-grit abrasives, and is usually intended as a first step to finer finishing. While polished finishes of #3 and higher have a grain pattern like brushed finishes, they are reflective, which brushed finishes are not. A very bright finish, more silver-colored than gray, can be found in level #4, with its defined brush pattern obtained using 120- to 150-grit abrasives. It is commonly specified for appliances, sinks, and hardware. Also having a slight brush, although more "satiny" in appearance than

#4 is #6. For a highly reflective surface with very slight brushing visible one would specify #7. The final designation, #8, is a fully specular finish with no visible grain. Sheets produced with finishes of #4 or higher are usually coated with plastic to protect them from scratches during shipping.

In our experience, linear brushing has been specified almost by default by architects and designers because of its predictable results. However, the linear quality of the pattern makes it fragile, since a single non-parallel scratch can ruin a part. Brushed and polished metal parts are very delicate, and a deep scratch can take many hours to grind off and blend into the finish, if this is even possible. On softer metals, such as brass and bronze, the finishes are easier to maintain, since the necessary grinding can be accomplished more quickly. For stainless steel, non-linear finishes are sometimes a better choice for areas subjected to heavy use. A particularly good finish can be obtained using a handheld random-orbital grinder (see Chapter 10). The random pattern camouflages many small scratches, and deep scratches can be hidden by regrinding at the job site, as the pattern produced with the tool blends easily.

ORGANIC COATINGS

Organic coatings are continuous coatings deposited on the surfaces of parts. They are used primarily for two reasons: to inhibit corrosion of base metals and to provide a decorative finish. Corrosion can be defined as the transformation of a metal from its pure state into a state of combination with other elements. For most metals, the contamination comes from exposure to the atmosphere and moisture, resulting in oxidation. On ferrous metals like steel, oxidation is commonly called rust. Rust destroys parts by weakening them and freezing them together.

The family of organic coatings includes paints, varnishes, enamels, and lacquers. "Painting" is a system composed of several steps beginning with surface preparation, the application of a primer, and finally the application of an organic coating. Paints are made from a number of components that include binders, solvents, and pigments. Varnishes are clear solutions of resin mixed with solvents made from oil or volatile liquids that evaporate, leaving behind a solid coating. Enamels are similar, but also have pigments added. A lacquer is defined as a solution of shellac, resin, cellulose, or polymers that is also mixed with an evaporating liquid solvent.

Binders form layers of protective film on a surface. The most common binders are drying oils, alkaloid varnishes, and synthetic resins like vinyl. Binders form solid layers when exposed to the atmosphere, but they lack hardness. To make paint more durable, they are mixed with a wide variety of resins. Paints are also mixed with substances called driers, which speed their rate of solidification. Paints and other organic coatings may include such special additives as silica (to create non-skid surfaces), anti-oxidants, and fire-retarding chemicals.

Some of the resins used for metal coating include alkyd resins, phenolic resins, polyurethane resins, vinyl resins, epoxy resins, and acrylic resins. Each is suited to a slightly different application. Alkyd resins are used in maintenance coatings because they are very hard and glossy. Phenolics are useful in waterproofing. Polyurethanes can produce coatings with superior gloss and color retention. Vinyl and epoxy resins are resistant to many chemicals and are highly weather resistant. Acrylic resins are durable and especially resistant to damage from ultraviolet (UV) light. Each resin requires a different solvent to keep the mixture free-flowing. Common solvents include alcohol, esters, ketones, and ethers.

The improper use and disposal of solvents causes environmental degradation, and regulations regarding their use are only growing more restrictive over time. As they dry, solvents release volatile organic compounds (VOCs) that combine with other pollutants to form ozone, which causes lung damage. There is often a gap between the "ideal" solvent in terms of performance, and one that can be legally used. It is important to check with local suppliers and painters to determine which paint should be used for a particular application.

Pigments not only provide color, but they also hide base material, inhibit rust, and protect from the ultraviolet radiation that can break down some coatings. There are numerous chemicals and organic compounds added to coatings to provide different colors. Zinc, aluminum, and lead dusts are frequently added to paints intended for metals, both for their color and various protective properties. Zinc provides galvanic protection, while aluminum blocks pores in the surface of the paint to help prevent water penetration. Lead reacts with the binder to form chemical "soaps" that inhibit corrosion.

SURFACE PREPARATION

For a coating to bond properly, the surface of the base metal must be completely clean and free of impurities. Poor surface preparation is frequently cited in paint technology literature as the primary cause of coating failures. There are a number of processes used for surface preparation, including solvent, acid or alkali washing, grinding, sandblasting (see Chapter 11), water blasting, and wire brushing. Aggressive finishing both removes impurities and roughens the surface of the metal, which gives the paint more surface area and deeper penetrations to improve its grip. It is particularly important to remove mill scale and rust patches from hot-rolled steel, as these patches are weak and can easily flake from the surface. For deep holes, the body filler used on automobiles can be used on bare metal parts to provide a rough surface for primer adhesion. It is only suitable for opaque coatings, however.

Phosphate coatings are also used on ferrous and some non-ferrous metals to improve the surface for painting. They are produced by dipping pieces in a solution containing acid phosphate salts, which cause tiny crystals of iron, zinc, and manganese phosphate to form. The crystals both deter further oxidation and create a microscopically rough surface to help bond paint. This kind of pretreatment is sometimes called an etch or a wash primer.

Stainless steels can be treated prior to painting with a process called passivation. Passivation is the process of coating stainless steel with a thin layer of oxide film to deter further corrosion. Actually, passivation is normally performed to *repair* that layer after it has been damaged; stainless steel already has such a layer, which gives it its excellent corrosion resistance. Situations that call for passivation include parts that have been cut or welded extensively. In these cases passivation is a good idea, even if the parts will not ultimately be painted.

Plating can also be thought of as a pretreatment for the base metal. Steel can be coated with a layer of molten zinc using a process called galvanizing (see Chapter 14). Paint does not adhere well to galvanizing, however, so the zinc must be additionally treated with etching or coated with special primers.

PRIMER

Primer coats are applied after pretreatment and prior to the surface coat. Different primers serve many purposes, including corrosion protection, covering slight surface irregularities, providing a highly bondable surface, and even creating a sandable layer that can be made smoother than the base metal.

For most metals, primers are made from alkyds, resins, and inorganic zinc. Alkyd primers are good for adhesion to surfaces with little or no pretreatment, but do not provide good corrosion suppression. Resins are used for pretreated surfaces and offer moderate to good protection when used in conjunction with a top coat. Zinc primers, when used over cleaned and roughened surfaces, offer the best protection. Most zinc primers can even be used without a top coat. Metal that has be galvanized must be primed for painting with special mixtures containing special latex emulsions, zinc dust, or Portland cement.

Since primer usually has a dull matte finish, an intermediate layer, called an undercoat, is placed over the primer prior to final painting. Undercoating is essentially decorative and serves to hide both the texture and the color of the primer layer.

CLEAR FINISHES

Designers often want to retain the appearance of the base metal while also obtaining protection from corrosion. Clearcoating can be achieved with varying degrees of success depending on the base metal. The most obvious problem with clearcoating is that an appropriate primer cannot be used. For this reason, metal should be thoroughly cleaned prior to clearcoating. The most common clear finishes used for steel are polyurethane-based, but in our experience it has not been uncommon to find fine webs of rust appearing over time as the paint develops tiny fissures. Rusty patinas on steel can also be preserved with polyurethane, but the coating will darken the finish, making it more brown than orange. If you are trying to preserve a patina on copper, brass, or bronze, there is a product known by the trademark name of INCRALAC™ that is specially formulated for this application. INCRALAC™ is an acrylic lacquer with special additives and a xylene-toluene solvent. Unlike polyurethane, it is well suited for metal in exterior applications, but xylene solvents are not legal in some jurisdictions.

As a general rule, it is probably better to leave metals (with the exception of carbon steel) exposed or coated with clear wax if their natural appearance is desired. Stainless steel is naturally resistant to corrosion, and copper, brass, and bronze form attractive oxide coatings that protect them and can be periodically removed if they are not desirable.

APPLYING ORGANIC COATINGS

Organic coatings can be sprayed, dipped, or applied with a brush or roller. Ideally, metal parts are coated in a controlled environment prior to jobsite delivery, since field conditions are always inferior in terms of dust contamination, overspray risk, and access to the work piece. At a painting facility, parts can be pretreated, primed, and given a top coat while they are carried on conveyors or racks. It is important to provide painters with some way to hold the parts during the process; a hole or place where the part can be clamped that will not need paint must be provided. This rule is generally true of painting, plating, galvanizing, anodizing, and passivating. Also, some paints are baked to achieve final curing, so assemblies must break down into small enough components that they will fit in a curing oven. While the curing oven is usually very large, it is best to check with your painter.

Painting parts in a spray booth using liquid paint with an airless spray gun.

Organic coatings that are sprayed are usually applied in a paint booth, an enclosed room with powerful air handling equipment. The air in the booth is filtered to trap overspray (paint that is sprayed but does not adhere to the part). Conventional paint sprayers use either compressed air or hydraulic pressure (called airless spraying) to atomize and throw the paint particles. With both processes, substantial amounts of overspray are produced, significantly worsening atmospheric pollution.

Air quality regulations have forced many manufacturers to abandon traditional organic coating processes in favor of a more advanced process known as electrostatic powder spraying. Electrostatic powder spraying is more commonly called powder coating. With powder coating, the part to be coated is grounded and a gun fires electrostatically charged dry paint particles in the general direction of the piece. The particles are attracted to the part and form an evenly coated layer on its surface. The charge particles are held on the part until it is taken to a curing oven, where heat is used to melt the powder. After heating, the liquefied powder solidifies into a hard, continuous coating. Powder coating does not require the use of solvents. The most common powders used for powder coating are epoxy resins, Teflon®, polyesters, polyvinyl chlorides, and nylon.

There are three critical limitations with powder coating. First, the piece to be coated must be made of an electrically conductive material (metals are excellent, wood and most plastics are not). Second, the base piece must be able to withstand temperatures of up to 400 degrees Fahrenheit in the oven while curing. This means

Applying powdercoat with a gun.

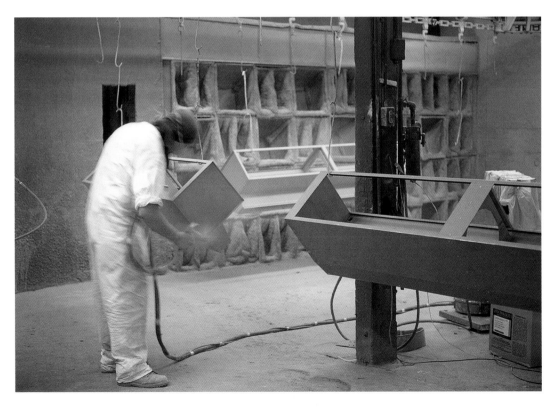

Powdercoating. A supply of dry powder (a), propelled by compressed air, is given an electric charge in the nose of a powdercoating gun (b) and sprayed onto the part (c) which is grounded, and attracts most of the powder.

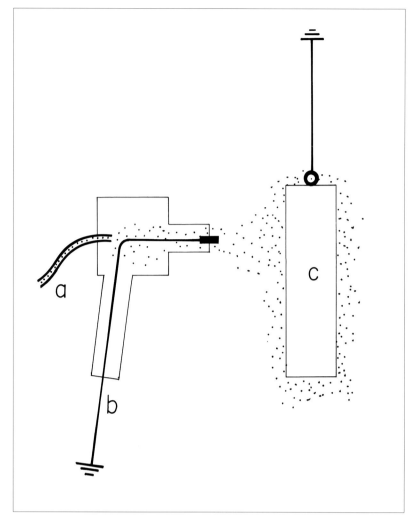

Racked powdercoated parts emerging from the oven.

that regular body filler products cannot be used to fill holes or smooth profiles; only relatively expensive, high temperature metal/epoxy fillers can be used. Third, the powders used cannot be mixed to create a variety of hues like most liquid paints; instead, a finite number of premixed colors are available.

The advantages of powder coating are its surface hardness, the ability of the charged powder to coat narrow crevices and undercuts, and the consistency of the colors and textures. Powder coating produces very little overspray, also adding to its cost-effectiveness. Given the trends in environmental regulation, powder coating seems poised to become the primary method of painting for metal products.

VITREOUS ENAMEL

Although vitreous enamel is not an organic coating, it provides another way of coating metal with a protective layer. Vitreous enamel is also known as porcelain enamel. It consists of a hard glass that is fused to the base metal and securely bonded. Crushed glass and pigment are mixed with water to form a paste, which is applied to the part cold. The coated piece is then heated (as hot as 1600 degrees Fahrenheit), causing the glass paste to melt into a smooth liquid layer that bonds firmly to the metal as it cools. The resulting finish is much harder than any organic coating, but it is also more expensive to apply. Vitreous enamel is commonly used for bathroom fixtures and architectural panels.

PLATING AND ANODIZING

Just as metal parts can be painted or powder-coated to protect them from corrosion and change their surface appearance, they can also be plated or anodized. Plating is the process of depositing a thin layer of metal over the surface of a part. Anodizing is a process (used primarily on aluminum) that alters the surface of a part to make it more resistant to corrosion. Decorative colors can also be added to the anodized layer and sealed.

PLATING

Plating is an alternative to building parts entirely from a metal whose appearance or physical properties are desired, at least superficially. The desired metal may be too expensive, too difficult to work, or too weak to use exclusively, so instead the part is made from a base metal and coated with a very thin layer of plated metal. Plating is divided into three major categories: electroplating, hot dipping, and spray metallizing.

Brass plating over polished bent steel tubing.

ELECTROPLATING

Electroplating is a technique used to coat parts with a different metal by means of electrolysis. Usually this is done by suspending a part in a tank containing a liquid acid or alkaline solution called an electrolyte. The part is given a negative electric charge, and a positively charged "anode," sometimes made of the plating metal, is also immersed in the solution. The electrical charges cause a flow in the electrolyte of particles of the plating metal to the base part, which quickly becomes coated. The anode, if it is made of the plating metal, disintegrates into the mix.

Two primary design criteria are evident from this description. First, the part to be plated must be small enough to fit into whatever tank the plater has available. Second, there needs to be some way to suspend the part so that it can be raised and lowered into the tank. Parts are usually suspended in groups on a rack or individually hoisted with cable or chain, but there has to be a hole in the part through which it can be secured. Consult with your plater as early as possible in your design process to confirm his requirements and limitations.

Plating. A negatively charged part (a) is suspended in a tank of metal salts (c) along with a positively charged anode (b), causing a flow of metal particles to the surface of the part.

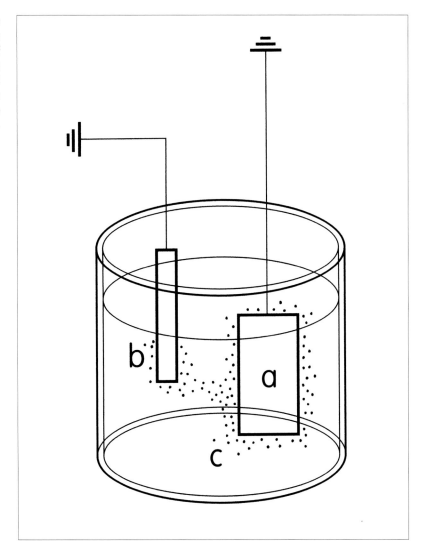

Another major design consideration is the "throwing power" of the plating solution. Depending on the shape of a part, plating metal will be deposited in different thicknesses at different positions on a work piece. Platers use chemicals like cyanide to improve throwing power, but some shapes pose unexpected challenges. Also, because the plating metal is expensive, it is undesirable to have parts that might trap solution inside voids, like the insides of tubes. Pieces with voids must be designed so that the solution flows completely out of a part as it is raised from a tank. This can usually be solved by drilling a few judiciously placed holes.

Parts to be plated are first cleaned thoroughly to remove oxidization, dirt, and grease that might interfere with the adhesion of the plating. Most metals can be plated, but since aluminum forms a natural layer of oxidation immediately upon exposure to air, it is difficult to plate. Electroplating usually leaves a microscopically thin layer of deposited metal, so the surface texture of the base part shows through clearly. If, for example, a polished finish is desired, the surface of the base part has to be polished before plating occurs. Very little work can be done to an electroplated surface without rubbing it away, since the plated surface is typically as thin as .002".

Nickel and chromium are two of the most common electroplating metals used. Zinc is also used, although it is more commonly applied through hot dipping. Other electroplating metals include copper and copper alloys, silver, and gold. Companies that provide plating will often also provide a wide variety of patina options, using acid solutions and heat to create desired colors. They will typically also

Racked steel parts ready for plating.

A plating facility. The operator is removing a rack of parts that have been plated.

The parts after plating with a very thin layer of brass.

provide mechanical finishing (brushing and polishing). Plated surfaces are sometimes sealed with clear paint or powder coat for added protection.

HOT DIPPING

Hot dipping is the process of immersing a part in a molten bath of plating metal that coats the part and bonds to its surface. The most common metal deposited by hot dipping is zinc, and the process is commonly referred to as galvanizing. Zinc protects the base metal (usually steel) by sealing it from the atmosphere to shield it from oxidation. Even if a zinc-plated surface is scratched, it will continue to protect the base metal by serving as a "sacrificial anode," allowing itself to be consumed first by corrosives (see Chapter 1).

Galvanized steel part.

While electrode-deposited finishes are very thin, hot-dipped finishes are comparatively thick. Like paint, hot-dipped plating is subject to drips and edge thickening. If a zinc-plated part is cooled slowly, it can form "spangles," large crystalline forms that resemble snowflakes. If it is cooled quickly, the crystals are too small to be seen and the finish is referred to as "matte."

Hot dipping is subject to the same handling criteria as electrodepositing; parts need holes to be hung or hoisted, and they must not be larger than the plater's tanks. Voids that do not drain are extremely dangerous with hot-dipped parts, as gas and molten zinc can become trapped and result in explosions. Also, the very high temperatures used can cause thin steel parts to warp. Hot dipping is not a delicate process; keep this in mind when specifying it.

SPRAY METALLIZING

Spray metallizing is the process of spraying a protective layer of molten metal on a base material. The metal is melted, atomized (reduced into tiny particles), and sprayed onto a surface using compressed air. Spray-metallized finishes are very grainy and somewhat uneven in thickness when applied by hand. What is extremely interesting about this process is the fact that it can be used over almost any substance, including materials like plastic, rubber, or even wood, since the metal cools almost immediately and does not significantly heat the base material. This process is useful for creating hard metal layers over continuous compound shapes that would be difficult or impossible to create using other fabrication methods.

ANODIZING

Anodizing is very much the opposite of plating. Instead of coating a base metal to protect it from oxidation, with anodizing the base metal itself is oxidized to create a protective layer. The protective layer is harder than the base metal, and therefore becomes resistant to further oxidation. Very frequently, color-producing chemicals and even dyes are added to the anodized layer to create decorative finishes, and also to colorcode parts in assemblies like those found in aircraft. Aluminum is the metal normally treated by anodizing.

The base part is given a mechanical finish (brushing and polishing), is degreased, and then immersed in an electrolytic bath. The most common electrolytes used are chromic and sulfuric acids. The part is given a positive electric charge, and a negatively charged cathode is immersed in the bath. The oxidized layer grows inward toward the center of the part; it is typically very thin. The layer is composed of a porous outer layer and a very thin, non-porous barrier layer that separates the oxidation from the base metal. Because the layer is very hard, it is difficult to cut or drill into parts after anodizing, so only completed parts should be taken through this process.

Anodized parts.

To produce color, an anodized part can be immersed in any number of aqueous salt solutions prior to sealing. These react with the porous layer to produce different colors. Color can also be added in the form of dyes, which become impregnated in the porous layer. The anodized part can then be sealed by boiling (which changes the size of the outermost aluminum molecules and closes the pores), by using a wide variety of chemical treatments, or even with wax or oil. The colors produced are typically very bright and translucent, giving them a slight metallic sheen.

With both anodizing and plating the available range of finishes will vary from vendor to vendor. Many processes executed in both finishes are proprietary, so it is worth your while to collect samples from different companies to compare their capabilities. Like many aspects of metal fabrication, research and experimentation continually result in new techniques and finishes.

STRATEGIES FOR CREATING VOLUMETRIC SOLIDS

There are two basic strategies for fabricating volumetric shapes in metal. The first is to use a skeletal frame to support a "skin" of surface panels. The second strategy is to weld together the surface panels themselves and rely on their rigidity and shape for dimensional stability. The former method is usually appropriate for larger constructions, while the latter tends to be applicable only to smaller objects.

FRAME-SUPPORTED SHAPES

For large objects (roughly speaking, anything larger than a table base), it is often appropriate to build a skeletal frame that approximates the desired shape to serve as an armature for the attachment and support of exterior panels. The reason for this is that an unsupported plate must be relatively thick when spanning long distances to prevent "oil canning" (surface deformation due to unequal shear stress), and the extra weight of pieces made from thick plate adds significant material and handling costs. A skeleton provides a measure of rigidity, even if the metal skin is very thin. As a rule of thumb, 16-gauge steel panels should not span more than two feet unsupported, and 11-gauge steel panels should not span more than about four feet if oil canning is not desired. These spans can be increased (sometimes substantially) if the panels are folded (bent) to create rigid membranes.

Armatures can be built from any metal extrusion, although lengths of tubing, channel, or angle are typically used. Straight lengths can be mitered to produce angular connections, and curved faces can be built with rolled extrusions. Metal sheet is often flexible enough that it does not need to be rolled to assume broad curves; it only needs to be secured to the curving armature.

A mid-century airplane fuselage made by applying a thin skin of metal panels to a skeletal armature.

A volumetric shape suggested by an armature made from thin round solid bar.

Metal panels can be secured to the frame using mechanical fasteners. If screws are used, they can either be backed up with nuts or screwed into tapped holes in the armature. Another strategy is to weld studs onto the frame, mating them to holes in the panels and fixing them with a nut exposed on the face of the panel. If the panels are to be painted in a shop or treated in some other way (galvanized, plated), it usually makes sense to make the panels removable, since the size of spray booths and plating tanks may be smaller than the entire assembly.

Panels can also be welded directly to the armature along their edge, from behind, or blind welded from the front. A blind (or rosette) weld is produced by drilling a hole in the surface of the panel, clamping it to the frame, and depositing welded material into the hole. The welded material fuses the panel and the frame

The steel skeleton of a building prior to the application of metal panels.

A series of hollow bases
made from 11-gauge
cold-rolled steel sheet,
continuously welded
and ground.

A series of hollow bases
made from 11-gauge
cold-rolled steel sheet,
continuously welded
and ground.

A series of hollow bases
made from 11-gauge
cold-rolled steel sheet,
continuously welded
and ground.

together, and can be ground smooth to make the joint undetectable from the panel side (see Chapter 8).

Welding panels directly to their frame is a fast assembly method, but keep in mind that welded panels are more or less permanently attached. This will pose a serious problem if an individual panel is damaged and requires replacement. On the other hand, fasteners by their very nature make panels susceptible to theft. Although there are many kinds of "theft resistant" screws that rely on unusual drivers for removal, anyone with enough determination can buy a matching driver.

If the designer's intention is to create the illusion of a unitary object, it is important to remember that the observer will readily perceive obvious panel joints and fasteners on close inspection. If it is desirable for the visual impact of the parts to yield to the whole, the designer would be wise to locate panel joints at planar intersections or at regular intervals that will not distract from the overall composition.

Architectural metal wall panel systems merit investigation, both for inspiration or application. Most prefabricated systems rely on clips and corresponding extrusions to create walls that are devoid of observable fasteners. Clips are a good way to attach panels, but remember that welds used to fasten a clip to the back of a metal sheet will telegraph some deformation to the front of the panel. The grinding and sanding required to conceal the telegraphed weld can add considerable cost to the job.

Many of the same issues that apply to the exterior cladding of buildings also apply to large sculptural objects when they are exposed to weather. Shapes should be designed to shed (and never pool) rain and snow so that water damage to surface finishes is minimized. Drain holes should be placed at the bottom of shapes to allow internal condensation to escape. Dissimilar metals should be separated by nonmetallic spacers or washers to prevent galvanic action (see Chapter 1).

Allowance for thermal expansion and contraction of metal panels is also a critical consideration in the detailing of large objects. A four-foot-wide steel panel could easily expand as much as $^1/_{16}''$ between the hottest and coldest days of the year. Expansion can be accommodated by using slotted fastener holes that allow the panel to stretch without building up internal stresses, and by providing space between individual panels. Many kinds of specialized expansion joints have been developed for buildings, and these products can be applied to any metal-framed object.

BUILT-UP SHAPES

Relatively small shapes can often be built without an internal armature by welding together the edges of the surface plates or sheets. Usually only steel and stainless steel can be used for this; softer metals like copper and brass, when used in sheet form, should always be backed up by an armature. Edges may be continuously welded and then ground smooth to create a seamless joint. It is typical to specify 16-gauge for small parts that will not be exposed to potential impact damage, and 11-gauge sheet for pieces that will see rougher service. One-quarter inch-thick plate is nearly indestructible, but it is also heavy and comes only in a hot-rolled form, which is not always desirable. Sheet that is thinner than 16-gauge will tend to deform under continuous welding.

The geometry of self-supporting fabricated shapes is limited by some basic rules of economy and material characteristics. First, it is inadvisable to design pieces with multiple curvatures (curving simultaneously in two directions on one plane). While these curves can certainly be formed by stamping, they are difficult to produce from flat stock using cold bending alone. Panels with multiple curves can be formed by heating and hammering them over shaped bucks, but this technique belongs more to the past world of blacksmithing than to modern fabrication. The skill necessary to do this well is no longer common, probably to our loss. An economical alternative for the formation of uniform complex curves is metal spinning

Volumetric solids that really are solid. These pieces were made by cutting sections of solid extrusions and layering them.

Volumetric solids that really are solid. These pieces were made by cutting sections of solid extrusions and layering them.

(see Chapter 6). Keep in mind that spun parts are usually made from thin sheet, so they will not hold up well under continuous welding. Second, it is very difficult to grind welds inside intersections, but very easy when they are exposed on the outside of a shape. It is a good idea to specify that inside corner joints be welded from the back, leaving a hairline crack on the outside. This is almost visually indistinguishable from a solid joint, although it requires extra planning on the part of the fabricator to sequence the welds. Another possibility is to make inside corners by bending, but this results in a rounded edge that can be more noticeable than a butt joint.

DETAIL AND VISUAL SCALE

This chapter provides guidelines any specifier of custom metalwork should keep in mind during the design process.

WORK WITH EXISTING PRODUCTS

- Become familiar with the existing palette of material products (see Chapter 2) and use them as the building blocks of your designs.

- Look carefully at the existing range of extrusions and sheet products. Note that many stock items, such as hot-rolled steel angles and bars, are not perfectly square in profile, and design with that in mind.

- Steel plates are far from perfectly flat. In order to make a precisely flat $3/4$" steel plate, it is necessary to start with a 1"-thick plate and use heavy grinders, called blanchard grinders, to remove enough stock to find truly flat surfaces.

- Unbraced extrusions and sheets do not remain straight and true, and many are never entirely straight. Obviously, the thicker the piece, the more rigid it will be, but this truism is counteracted somewhat by the additional weight of thicker pieces, which can also generate sag. For structural considerations, an engineer should be consulted. For decorative work, it is prudent to design pieces to assume naturally rigid shapes, if rigidity is a priority.

- If the basic products in their unadulterated states do not meet the requirements of an element in your design, consider employing simple modifications. Tapers can be cut from extrusions, using any number of processes. Special angles and tees can be created by splitting standard I beams and channels. Custom extrusions can also be ordered (for a price) in fairly small batches in aluminum and cold-rolled bar. Custom-sized sheets can be procured directly from a rolling mill, but only for orders of many thousands of pounds. Before pursuing this route, consider the lead time necessary for such a specialized product.

- What if you discover during construction that you will actually need a few more sheets of a custom-sheet size, or a few more feet of a custom shape? Standard components are usually stocked in all common sizes in major metropolitan areas, so getting extras is not calamitous in terms of time or budget.

**EMPLOY
PRECISION
ONLY WHEN
NECESSARY**

- Recognize that the precision available with certain metalworking processes can be pointless if it is not matched by equal precision in the materials with which the metal will interface. It is hubris to expect accuracy on a construction job site greater than $1/8"$, let alone $1/16"$. Thus, the tolerances obtainable on a mill of .001" or less are rarely applicable to anything other than machinery. There is a tremendous cost differential between a dowel that is specified to be 12" ($+/- 1/16"$) and what is basically the same dowel, specified to 12" ($+/-.001$). The obsession in contemporary construction with obtaining precisely level and true horizontal and vertical alignments comes at a price.

- While precise squareness and flatness can be obtained in assemblies, it is not inevitable. When sheets and extrusions are heated by

A stainless steel library ladder, made from intricate stainless shapes cut from plate.

welding and cutting, stresses are created and "unlocked" which can
warp and twist parts, sometimes dramatically. For large production
runs, prototypes can be produced so that these stresses are
anticipated and counteracted, either by tightly clamping parts at
key points or by carefully planning the order in which welds are
placed. When only one or two editions of an assembly are being
made, the fabricator must rely on his experience to determine
where clamps will be placed or the sequence of welding.

• If the design of an assembly is such that slight deformations will
detract from the composition (for example, on a large steel cube
made from sheet, or a very simple table with a clear transition
from the horizontal top to the vertical leg), the costs of fabrication
will substantially exceed those for less rigorous detailing.

- In architecture, moldings are the primary device used to conceal imperfect construction. Moldings are ornamental strips placed at joints inside and outside buildings. Whatever their aesthetic function, their real purpose is to distract the eye from the critical intersection between walls, floors, and ceilings, which inevitably bow and lean. Consider the use of metal "moldings" at joints.

- Three general rules to consider regarding accuracy when designing for fabrication:
    1. Design elements with tolerance at interfaces. This means oversized holes and limited connection points.
    2. Design elements that can be adjusted to stretch or contract slightly to accommodate jobsite variations.
    3. Explore joints that allow overlap or express their adjustability, rather than those that demand precise alignment.

LARGE-SCALE ENGINEERING SOLUTIONS ARE OFTEN SCALABLE

- Industrial and utilitarian structures made from metal components are a rich source of inspiration for systems of assembly. Because metal components are produced in a wide variety of sizes, systems used to assemble large structures can often be scaled down with more or less the same extrusions for smaller assemblies. Look to these for creative design solutions.

- Utilitarian structures (billboards, mine shafts, bridges, water towers) offer unique compositional strategies for resolving unusual load conditions.

The armature supporting a billboard.

A flip-chart stand
designed with billboard
imagery in mind.

CONSIDER HOW THINGS THAT ARE TOUCHED DIFFER FROM THINGS THAT ARE VIEWED

- Something small enough to be touched differs in detail from elements seen from a distance. While a tubing intersection on a handrail might seem crude if the weld is left unground, on a larger element, such as a cement hopper, the weld point is not visually troubling. We demand a far higher standard of consistency from objects experienced with tactile senses than from those that are only observed.

- Polished surfaces on things that are touched are both pleasing and functional. They can be maintained by repolishing, while coatings such as paint rub away.

- Consider knurling on items that must be securely gripped.

- Remember that a sharp edge on a piece of exposed metalwork can be dangerous. Edges must be eased, and corners must be slightly broken.

- The level of detail you develop in a group of elements reflects your vision of metalworking. Very rough work, such as plasma-cut, hot-rolled steel plates with unground stick welds, constitutes a sort of self-contained language with its own set of rules. This is a language that can be used by a designer to express a way of identifying with the material that is entirely different from that of the designer who expresses his ideas in carefully machined, highly polished metal components. It is unlikely that a designer specifying hot-rolled steel would be offended by the fabricator leaving stampings from the mill, or even chalk marks added during construction, on the face of the finished piece. Leaving such marks would be inconceivable to the second designer.

- Since the fabricator must ultimately be trusted to make certain decisions that will correspond to the designer's vision, the specifier needs to be very clear in expressing his sensibilities.

A decorative fixture made from hot-rolled steel plate, shear cut, with unground welds.

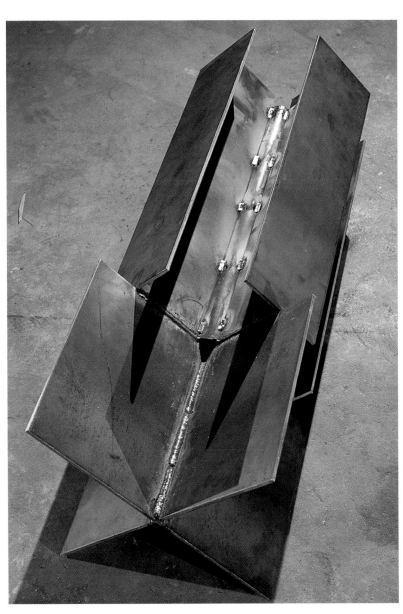

CHAPTER 17

MOTION

Moving objects are exposed to a wide array of repetitive and unpredictable forces. While many materials are strong in either tension, compression, or shear, few materials other than metals possess strength in all three loading conditions. For this reason, moving objects are frequently reinforced at critical joints or made entirely from metal components. What follows is a brief discussion of the basic mechanisms that can be used to transmit and regulate motion.

Motion is imparted by an external force acting on a stationary object. The force has direction, called vector, which accelerates the stationary object away from the force. In man-made devices, force can be generated by harnessing natural sources such as gravity, wind, rushing water, or sunlight. Force can also be generated by engines, which burn fuels, or motors, which translate electrical energy into motion. Motors and engines usually direct their motion to a rotating rod called a shaft. A stationary rod, on which a moving object rotates (such as a wheel), is called an axle. Most mechanisms incorporate either a shaft or an axle in their design.

Moving objects will usually have one or more intersections where stationary parts come into contact with moving parts. This joint can occur at the shaft, the axle, or at another point where surfaces slide against each other. Depending on the smoothness of these surfaces, varying amounts of friction are generated by their rubbing. The friction causes heat, tearing, and shearing of material, which may eventually inhibit motion.

It is critical to use materials at the motion interface that are extremely smooth, or even better, to isolate the surfaces from each other with the use of a lubricant. Oil is probably the most common lubricant used in metal assemblies, since it coats parts evenly, remains in a more or less liquid (and therefore frictionless) state, and has the added benefit of shielding metal parts from oxidation. Cushions of air are used in some situations to create frictionless layers, particularly on work tables. Silicon is another common lubricant. Using Teflon® is often a simple way to create smooth motion interfaces since it is virtually frictionless. It can be purchased in sheets, tubes, rods, bars, tape, and rings for various applications. Soft metals such as bronze can also create smooth joints. Bronze rings are available that come already impregnated with oil. When the ring is heated by friction, the lubricating oil beads to the surface.

Another approach is to minimize the size of the contact between the faces by using wheels and balls. Owing to their geometry, only one point on a ball or wheel touches an opposing surface at any time. In machinery, hard, precisely ground metal balls are set into lubricant-filled raceways or rings that allow them to freely roll against an inner and outer surface. These devices are called ball bearings. When formed into rings they are designed to fit onto stationary axles or rotating shafts. On a shaft, the outer ring is held stationary by compression or by a set screw, while the inner ring, in tight contact with the shaft, spins freely. On an axle, the relationship is reversed. Bearings can be spherical or barrel (wheel) shaped, and they are either slightly exposed to allow for easy lubrication, or sealed to protect against dirt and other friction-producing contaminants.

Bearings are often placed in housings called pillow blocks, which are machined steel blocks with mounting holes that can be secured with screws. Pillow blocks are meant for use in pairs because the bearing can be swiveled on its axis inside the block to adjust the shaft angle. In a pair, one locks the other into position. Another common use for ball bearings is in rotating tables, commonly known as "Lazy Susans." Instead of accommodating turning along the edges, rotating tables are designed so that a top and bottom plate spin freely over a layer of balls in a raceway. Like pillow blocks, they usually have holes for mounting to assemblies.

Wheels are a very common motion-transmitting device. Most wheels used in furniture and machinery are attached individually to bearings, axles, and mounting plates, and the entire assembly

Bearings.

145

Pillow blocks.

Casters.

is called a caster. Casters have wheels made of materials such as neoprene, plastic, steel, and wood, and are either fixed to roll in a single direction or mounted to yet another set of bearings that allow them to swivel. Casters can be supplied with simple foot-activated brakes, which compress the wheel to prevent rolling. Ball casters use a single ball instead of a wheel to transmit motion. The ball is mounted into a lubricated cup and held in place by exposing slightly less than half of the sphere.

For many structures the unadulterated rotation of a shaft is the only motion needed. Many machines, such as fans, are that simple. Shafts can also be fitted with flexible couplings, to allow them to turn another shaft rotating at an angle. The most versatile coupling of this type is called a universal joint. Universal joints permit the attached shaft to spin in virtually any direction.

With more complicated devices, it may be necessary to adjust either the direction or the speed of the motion from the shaft. The basic mechanism used for this purpose is a gear. Gears are circular wheels with teeth at their edges that interlock with teeth on other gears, racks, or chains to transmit force. They have one hole at their center used for mounting to a shaft or an axle.

Spur gears are star-shaped and have machined teeth on their edge. They are often matched with other gears of larger or smaller size in order to vary the speed of rotation. Spur gears can also be rolled against a toothed, linear rack that creates a raceway for the gear to follow. When attached to chains, gears transmit force in line over large distances. This kind of gear is called a sprocket.

Chain on a sprocket.

A spur gear
on a rack.

Pulley.

Spiral bevel gears.

Belts can be placed tightly around grooved wheels, called pulleys, to accomplish the same result. Some gears have beveled teeth cut on their faces that radiate from the shaft hole. These gears are known as spiral bevel gears, and are used to transmit force at right angles. Another type of gear, called a worm gear, is rotated by a spinning threaded shaft (called a worm), which is mounted perpendicularly to the gear. The threads on the worm engage the teeth on the edge of the gear. Worm gears are used for the transmission of power when the desired rotation is in one direction only. Reversing the worm will cause the teeth and the thread to seize. Worm gears are different from all other gears, which rely on other kinds of mechanisms to limit the direction of their rotation. A special gear, called a ratchet, has its direction controlled by a small pin called a pawl. The pawl is curved to bounce out of the way of the ratchet teeth when the ratchet turns in one direction, but lock forcefully into the teeth

Worm gear.

Ratchet and pawl.

when it turns in the opposite direction, thereby jamming any further motion. Ratchets are usually attached to a spring that returns them to their original position after each pass of a tooth.

The teeth on a gear transmit motion much like the threads of a screw, and there is actually a type of threaded rod, called Acme thread, which is made specifically for motion transmission. Acme threads are much more deeply cut than screw threads, with far fewer threads per inch. A matching nut serves as the point of contact between moving parts and is used to create a spiraling motion along the major axis of the rod. Another threaded shaft device called a ball screw is threaded with rounded indentations so that its motion contact is with a set of freely spinning balls mounted into a special nut through which the shaft travels. The balls generate less friction than the rubbing threads of an Acme rod, and thus less energy is misdirected into heat.

Aside from rotating shafts, the other primary mechanical method of supplying motion is with the use of a piston. A piston is a shaft that is pushed in a cylinder by either compressed gas or force transmitted by a fluid (hydraulic force). Hydraulic piston assemblies rely on the incompressible nature of fluids to transmit tremendous forces through a pump to the shaft. They are used for extremely heavy load movement. Compressed gas pistons are for lighter applications; they store force in compressed gas, which pushes the piston when the gas is allowed to expand. The most common example of a hydraulic piston is the hand jack used to lift a car to change a flat tire. Gas pistons are often used as closers on doors.

A spring is an elastic device, usually a spiral coil of wire or flat bar, that returns to its original shape when released from the force of being pulled, pushed, or twisted. Springs absorb and repel, but do not transmit, forces. The three most basic spring types are compression, extension, and torsion. Compression springs are meant to absorb compressive forces, or pushing. Buttons that spring back into position after being pushed are mounted on compressive springs. Extension springs allow a user to pull an object and have it return to its stationary position when released. Handles on many machines are fitted with this kind of spring. Torsion springs resist rotational torque and are often found on self-closing door hinges.

Acme thread and a nut.

Hydraulic piston on a lift.

Springs are also used as an anti-loosening device with fasteners called lock washers. Lock washers are single-spiral springs that resist the vibrational unscrewing of fasteners by storing extra torque when they are compressed (see Chapter 9).

A compressed gas cylinder with a piston can also function as a spring. When the force is applied, the piston enters the cylinder, and when the force is released, the compressed gas pushes the piston back into position. A shock absorber is created by combining a gas spring and a mechanical spring; the mechanical spring absorbs smaller vibrations and the gas spring dampens larger motions. Very crude dampening devices can also be made from rubber or neoprene pads, which isolate machinery from vibrations transmitted at their points of attachment by compressing and expanding like a coil spring.

In architectural applications, hinges are one of the most basic categories of motion-enabling devices. Most hinges are based on a solid round pin structured like an axle, around which a tubular outer housing rotates. The housing is usually made of a series of alternating tubes stacked in alignment and attached to opposing leaves like an open book. The leaves are then fastened to the two parts of the system, much like a door and its frame. When these hinges are produced in long strips, they are called piano hinges. Most hinges are designed for mechanical fastening, but for metal objects there is a wide variety of pin hinges that can be welded. In general, door hardware, including closing, lifting, and locking systems, can provide a designer with a wealth of carefully engineered mechanisms that can be pressed into service for any moving element.

MOVEMENT OF HEAVY PIECES

The costs of shipping, moving, and installing heavy assemblies can constitute a significant percentage of the total cost of a finished piece. It is a good idea to try to design larger pieces so that they can be disassembled into parts that can be easily lifted by one or two people, and then reassembled at the jobsite by welding or bolting. Obviously, anything can be lifted by a large enough group, but the liability posed by doing this is not appropriate for most situations. It is also wise to walk the path an assembly will travel, from the point where it will be dropped off to its installed position. By doing this, the designer can anticipate clearance problems, especially at doors, tight corridors, elevators, and (worst of all) stairways. What follows is a description of the methods used to move heavy pieces from the shop to the jobsite.

When unassembled parts arrive or leave a fabrication shop, they are usually bundled for shipping on a basic platform called a pallet. A pallet is a sturdy base made from wood slats (usually 2 × 4s) arranged horizontally on two layers with a few vertical ones as spacers. Pieces can be tied, taped, wrapped with plastic film, bolted, or even glued to securely fasten them to a pallet. Pallets are sometimes used as a base for building a solid box around something to protect it, but usually parts are simply left exposed. When something is loaded on a pallet, it is described as being "palletized."

Pallets can be moved in several ways. They can be rolled around the work floor with a pallet jack, a hand-operated rolling dolly with a pair of reinforced arms inserted in a space between the top and bottom of a pallet. Pallet jacks have a short hydraulic piston that allows the user to lift the arms a few inches after insertion, thereby lifting the pallet off the floor. The jack has wheels that protrude from the bottom of the arms and a handle used for steering. Pallets can also be moved with a forklift. A forklift is a stout, highly maneuverable vehicle with a pair of "forks" at the front that are inserted like the arms of the pallet jack. On a forklift, however, the forks can be angled and also raised very high. The forklift is very heavy to help prevent overturning, and the driver sits in a protective steel cage. Most forklifts are run on propane, which is stored in a tank behind the driver. Smaller forklifts designed for warehouse use

can typically lift between 3,000- and 6,000-pound loads ten to
fifteen feet in the air. Construction forklifts are built to lift objects
weighing up to 6,000 pounds as high as 40 feet in the air, and are
equipped with sturdy tires for rough terrain. A heavy-duty forklift,
called a reach, can be used to lift objects as heavy as 10,000 pounds.

Palletized and rigid unpalletized items can also be lifted with
cranes and pulleys. Normally used for lifting engine blocks, small
cranes are used in some shops for routine floor-to-worktable lifting.
At jobsites, cranes are designed to lift objects from 10,000 to
150,000 pounds, although extremely heavy loads can only be lifted
limited distances. Cranes are expensive to rent, and any installation
requiring a crane should be carefully planned. Another common type
of crane in fabrication shops is called a gantry, which consists of a
hoist (a suspended gear or a pulley over which chain or rope is run)
attached to a trolley that runs along the bottom of a fixed I beam.

Hoists rely on mechanical advantage (see Chapter 5). On a hoist, the gear over which the chain is looped provides leverage proportional to its diameter. Simply stated, the bigger the diameter, the greater the leverage. It is interesting to note that the hook placed at the end of the chain on a hoist is designed to be weaker than the chain; the first indication of overloading is a straightening hook. Sometimes, the beam used for the hoist is a structural member in the ceiling. Other times, the beam is mounted to a pair of tripods so it can be positioned over whatever is being lifted. Hoists and trolleys can be either manually or electrically controlled.

On the shop floor, parts that are in process are often moved horizontally on dollies instead of pallets. A dolly is a low, sturdy platform with four swivel casters that allow it to be moved with ease. Specialized dollies exist with hydraulic and ratcheting platforms, both of which allow the user to lift loads to table level.

Lifting heavy plate with a hoist.

Roller track.

Vacuum lifter.

For repetitive processes, parts are moved horizontally on mechanized beltways or, more commonly, on roller tracks. Roller tracks consist of parallel runs of rolling cylinders that provide pieces with a frictionless surface on which to travel. Roller tracks are often placed in front of cutting tools so that heavy extrusions can be guided into position without lifting them.

While heavy pieces are usually lifted with forklifts or cranes, heavy sheets or plates (with a smooth surface) can be raised with a vacuum lifter, which operates by attaching a series of suction cups to the top of the plate to hold it tightly while it is lifted into place. Vacuum lifters are useful when large quantities of sheet must be moved one at a time from a pallet to a table or machine.

Once a piece is completed and has been palletized or crated, it must be placed on the shipping vehicle. Usually, pieces are picked up from shops with flatbed trucks. Flatbed trucks range from 12' to 24' in length and are about 7'-6" wide. For extremely long or heavy objects, larger trucks are available. Ideally, a shop will have an elevated loading dock that is level with the standard height of truck beds (about 48") so that pallets can be rolled right onto the truck. Otherwise, the pallets must be loaded with forklifts or by using a "lift gate" attached to the truck. A lift gate is a hydraulically operated platform that is raised and lowered by controls at the back of the truck. At the jobsite, the process is reversed.

Flat bed truck.

COMPUTER AND MACHINERY INTERFACES

It is hard not register a certain amount of surprise when entering a large fabrication facility for the first time. Brimming with expensive machinery designed for the production of very large and complex metal parts, the facility has only a handful of people working on the floor. Even twenty-five years ago, each machine would have been run by one or more specialized operators. What has changed is the application of computer numerical control (CNC) to a wide variety of machine tools that were previously operated by hand.

COMPUTER NUMERICAL CONTROL

Computer Numerical Control is the use of computer programs to control the movement of cutting tools, stock, the application of coolant, and a host of other processes. CNC has been successfully applied to mills, lathes, cutting systems (plasma, water jet, and laser), tube bending, and punching and stamping machinery. The use of computers, and the speed they add to fabrication, have made it possible to build larger and more diverse machines capable of accommodating a wide variety of processes in a single device. Accuracy that was virtually impossible to produce manually can now be obtained for critical assemblies. Tools can be changed automatically, and because sensors provide information regarding temperature and feed rate, subtle adjustments can be made in the cutting speed of tools, significantly increasing their life span.

The CNC control (a computer) interprets an NC (numerical control) program. NC programs provide instructions to the motors controlling motion on the machine. Usually, this motion consists of travel in the x,y, and z directions, but can also include information for rotation and tilting. The inclusion of tilt and rotation makes such devices "five-axis" machines. Most pieces of cutting equipment, such as laser and water jet cutters, function on an x-y grid only, since they are used for cutting shapes from flat sheet or plate. However, mill and lathe operations are frequently three-dimensional.

NC control on
a lathe.

NC programs are often written in what is called G code. G code programs consist of a series of data blocks designated by letter and number codes, each of which indicates a single operation to be performed (move in the cutting tool in the x-direction ___ inches, or set the cutting tool spindle speed at ___ rpm, for example). APT programming (which stands for automatically programmed tool) is another commonly used programming language that differs from G code in its use of descriptive words to indicate actions to be taken by a machine.

Many higher-end computer aided drafting programs (CAD) used by designers to create working drawings can create files that can be translated into NC programs and fed more or less directly to machinery for fabrication. Some CAD programs allow designers to watch simulations of the tools cutting the desired part in order to check for errors. Otherwise, a test part should be made on the actual machines using soft metal or a product like machinable wax.

If your CAD program is not capable of producing NC programs, you can usually send the basic information to your fabricator in the form of a DXF or other transfer file. A transfer file is written in a sort of "universal" format for the exchange of drawings between different CAD programs, and almost all CAD programs will translate drawings to it. Sending a transfer file gives the fabricator a set of measurements that he may be able to insert directly into his NC system, or at least can use as a reference to build an NC program for his machines, thereby saving time and money.

NC programs can be used to control machine operations on a point-to-point basis (punching, drilling) or to determine continuous paths that tools will follow to cut out contours. Continuous path control can be used to create extremely complex profiles in two or three dimensions. Advances in three-dimensional control could result in significant cost savings in the manufacture of precisely matched dies for press work (see Chapter 6), which at present are very expensive. A third kind of control, adaptive control, relies on sensors to report actual conditions at the work zone, such as overheating or dulled tools, so the computer can use this feedback to make continuous adjustments to the the work.

NESTING AND BENDING

Nested parts.

Another useful application of computers is their ability to generate nesting programs for cutting material. A nesting program will determine the most efficient layout of a group of parts to be cut from a sheet so that the least waste is produced. In the past, this

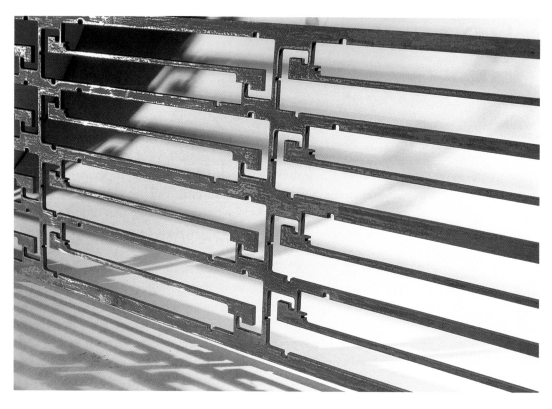

kind of layout could only be done by trial and error. On large jobs the savings offered by nesting can be substantial.

Making a bend in a metal part causes the material on the inside of the bend to shorten (due to compression) while the material on the outside lengthens (due to tension). Most excitingly, new processes using computers tend to rapidly decrease in cost, while their power increases. CAD software and a personal computer provide designers with entrée to a host of fabrication processes that were, until recently, beyond their reach.

DESIGN IMPLICATIONS

Although CNC machining and cutting is not inexpensive, it is capable of producing objects from metal with a level of accuracy that cannot be achieved in any other way. The obvious implication is that a new vocabulary of extremely complex shapes and patterns until recently beyond the reach of even the most skilled craftsmen is now available to designers. Intricate metal traceries that would be virtually impossible to produce without computers can be made, and then revised and customized with minimal effort.

The impact of computers will undoubtedly be as great as that of the harnessing of electricity was in the last century. While large commercial interests are what primarily drives technological advancement, the opportunities posed by change can benefit many sectors. For example, some familiarity with CAD and a personal computer provides a designer with entree to a host of fabrication processes that were, until recently, basically beyond their reach.

Very small, accurate parts made from sheet with a CNC-guided laser cutter.

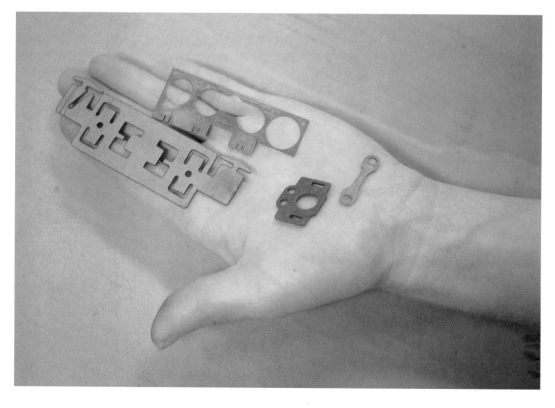

As job shops become increasingly attuned to producing small batches of parts in a climate of global price competition, it becomes more and more possible to incorporate custom items into architectural projects, or to produce short runs of manufactured products through processes that might previously have been accessible only to large companies. The ability of manufacturers to do this is in no small part the result of computerization, which eliminates the need for a variety of expensive jigs, patterns, and extensive setup time from a number of processes. Obviously, not every manufactured item can be produced cost-effectively in small quantities, but opportunities abound for custom production of furniture, architectural elements, and consumer products. Designers will find themselves greatly empowered by developing an understanding of how computers can be used to interface directly with manufacturers.

GLOSSARY
OF TERMS

ABRASIVES: Extremely hard, (usually crystalline) particles used to cut or grind metal. Abrasive particles are often bonded with resin to create grinding stones, or to resin and paper backing to create abrasive discs and belts.

ABRASIVE CHOP SAW: A cutting tool for extrusions that uses consumable resin discs impregnated with abrasive particles.

ACME THREAD: A thread size specially designed for motion transmission. For additional durability, the threads are thicker than those found on screws.

AIR COMPRESSOR: A device that stores compressed air in a metal tank for use as a power source on a variety of tools, including die grinders, sanders, paint sprayers, and plasma cutters.

ALLOY: A combination of a base metal with traces of other metals added to provide or enhance desirable qualities not found in the base metal alone.

ANGLE: An L-shaped extrusion with a cross section consisting of two perpendicular legs.

ANNEALING: Heating and slowly cooling a metal part to produce a crystalline structure that is softer and easier to work.

ANODE: The positive electrode in electrolysis. *See* electroplating.

ANODIZING: An electrolytic process used to create a hard protective layer of oxidation on the surface of metal (usually aluminum). Coloration can also be added to the surface of the part during the process using a variety of chemicals and dyes.

ARMATURE: A skeletal frame used to support a shape and provide attachment points for cladding with a "skin" of metal sheet.

AXLE: A rod that holds a spinning wheel or gear.

BALL BEARING: A raceway filled with spherical balls or barrel-shaped rollers that allows for minimal friction between a moving and a stationary part. When the raceway is ring-shaped, it is called a radial bearing, and is intended for use with an axle or a shaft.

BALL BURNISHING: A metal finishing process in which parts are placed in a spinning barrel filled with steel balls and tumbled.

BAND SAW: A cutting tool that uses a continuous cutting belt with serrated teeth. The belt is held taut by spinning wheels and run past a small opening in a flat steel plate, onto which the part to be cut is rested as it is fed into the blade.

BAR: Solid extrusions of metal, which in cross section have square, round, half round, hexagonal, oval, or rectangular (flat) cross sections. The availability of different bar shapes varies for different metals.

BEAD BLASTING: The use of small (usually glass) particles propelled by compressed air to condition the surface of a metal part.

BELT SANDER: One of a number of tools that spins abrasive belts to grind and brush metal parts.

BINDER: A material in an organic coating that dries to form a protective solid layer over the coated part. Drying oils and synthetic resins such as vinyl are common examples.

BITS (DRILL): Fluted shafts with cutting edges used in drills to produce circular holes. The spiraling flutes serve to carry chips from the cut zone.

BLACKSMITHING: The combined vocabulary of processes developed over millennia to work metal (in particular iron, which was referred to as the "black metal") in the absence of electrical power sources and stock metal products.

BLANK (SPINNING): A flat circular disc cut from thin sheet that is used on a spinning lathe to create shapes with compound curvature (e.g., half spheres, cones).

BOLT: Threaded fasteners intended for structural applications. During installation, the bolt is held steady while a nut is screwed onto it.

BRAKE: A tool used to produce bends in metal sheet and plate.

BROACH: A serrated, tapering tool that is inserted into round holes and pulled or pushed through to create a new profile in the hole (usually square or hexagonal).

BRUSHING: The application of even, linear scoring to the surface of a metal part.

BULGING: A process used to alter the profile of tubular parts. A "bladder" filled with liquid is placed in a tube and subjected to extreme force from a hydraulic ram. The tube, which is inside of a mandrel in the shape of the desired part, is forced against the mandrel and assumes its shape.

BUR: A small rotary grinding tool scored with serrated, hardened teeth. Burs are made in a variety of shapes for different grinding conditions.

BURRS: Sharp metal tabs that remain attached at the point of cutting or drilling operations. Burrs are undesirable; a variety of processes exist to remove them efficiently.

BUTT JOINT: A joint created by laying two plates edge by edge.

CAP SCREW: A high-strength screw with a hexagonal head. A socket head cap screw has a circular head with a hexagonal indentation for turning with a hex wrench.

CARBIDE: A hard metal often used in combination with other materials for cutting tools.

CASE HARDENING: The process of hardening the outer layer of a steel part by causing it to absorb traces of various materials (often carbon) and then heating the part.

CASTER: A component that consists of a wheel, an axle, forks, and a method for attachment (usually a threaded stem or a plate with holes). Some casters also have foot-activated brakes that work by binding the wheel when depressed.

CASTING: The process of pouring molten metal into a hollow mold, where the metal solidifies into the shape of the mold. There are numerous variations of the process available for metalwork.

CATHODE: The negative electrode in electrolysis. *See* electroplating.

CENTERS (LATHE), DEAD AND LIVE: Hardened rods ground to a point and located on a lathe at either end of a workpiece. Dead centers are stationary, while live centers can spin with the workpiece.

CHANNEL: A C-shaped extrusion with a cross section consisting of two parallel legs (flanges) and a perpendicular web.

CHROMIUM: A metal sometimes added to alloy steels to produce additional hardness.

CHUCK (DRILL): A device that clamps drill bits tightly with a set of adjustable jaws.

CHUCK (LATHE): A device used to securely hold a workpiece on the lathe.

CLAMP: A handheld, adjustable device used to securely hold unassembled parts together.

COLD-ROLLED STEEL: Steel products rolled cold into sheet and extrusions. Cold-rolled steel is more dimensionally accurate than hot-rolled steel and does not have a layer of mill scale on its surface.

COLD SAW: A cutting tool, similar in scale to an abrasive disc saw. A cold saw uses hardened steel cutting blades with teeth rather than abrasive cutting discs. The blade and the workpiece are kept cool by a stream of water-soluble oil. Cold saws produce clean, burr-free cuts.

COMPUTER NUMERICAL CONTROL (CNC): The use of computer programs to control and coordinate a variety of shop operations automatically.

COR-TEN STEEL: A high-strength, low alloy (HSLA) steel frequently specified for architectural applications. It forms a protective oxide layer when exposed to the atmosphere.

COUNTER BORES: A tool that produces a flat-bottomed, recessed "seat" at the top of a hole.

COUNTERSINK: A tool that produces a sloping recess at the top of a hole; the angle of the recess matches that of flat head screws so that they can be inserted into the hole flush.

DEBURRING: The process of removing sharp burrs from the cut edges of metal parts.

DEEP DRAWING: A press work process in which a hydraulic ram forces a "male" die into flat stock; the part deforms into a matching "female" die on the opposite side.

DIAMOND PLATE: A plate product produced by rolling a crisscrossing "diamond" pattern into the surface of one side. The product is used to provide slip resistance in such applications as stair treads and ramps.

DIE (PRESS WORK): A "master," usually made from hardened steel, that is pressed into flat material to cause it to conform to the die shape. A solid die is called "male," and a die that contains a void in the shape of the desired part is called "female."

DIE (THREADED ROD): A serrated cutting tool used to cut threads into the outer surface of round extrusions. The die is usually cylindrical, and is secured in a two-handled die wrench.

DISC SANDER: A tool consisting of a spinning circular plate mounted above a flat work table. Large abrasive discs are mounted to the front of the plate, and work is fed into the disc while resting it on the table. The angle of the table in relation to the disc can be adjusted to create beveled grinds.

DRILL, DRILL PRESS: A tool that spins drill bits to produce round holes. Handheld drills are fed into the part manually, while floor- or bench-mounted drill presses have a lever that is pulled to feed the drill bit into the part. The part is secured to a flat plate beneath the drill.

EDGE BANDING: The addition of a thin ribbon-like strip of metal, welded along the perimeter of a flat part, which is used to create the illusion of thickness.

ELECTRIC ARC WELDING (MIG, TIG, STICK (SMAW)): A variety of processes using electricity to generate the heat necessary to melt and attach separate metal parts. MIG welding uses spooled rod as both an electrode and filler material; TIG uses a tungsten electrode and separate filler rods; Stick uses a rigid electrode (sometimes called a sparkler) coated with filler.

ELECTROPLATING: A process for coating a metal part with a thin layer of another metal using electrolysis. A part made from a base metal is given a negative electric charge (making it a cathode) and immersed in an electrolytic bath containing positively charged ions of the coating metal. The coating metal is attracted to the base part and coats its surface.

EMBOSSING: A press work process in which patterns are stamped into the surface of thin sheet.

END MILL: A cutting bit typically used on a vertical mill and fed down into the workpiece to create slots and cuts along edges.

ENGRAVING: The process of cutting patterns into the surface of a metal part; this can be accomplished by using a mill or router with cutting bits or by chemical processes that etch the surface.

EXPANDED METAL: A product made from sheet by slitting and stretching a plate to create a three-dimensional pattern.

EXTRUSION: Linear metal products formed by forcing metal through dies to create a bar with a consistent cross section (see angle, bar, channel, I beam, tees, and zees).

FABRICATION: The process of creating assemblies from stock components.

FASTENER: A device or set of devices used to securely fasten two or more parts. Examples include screws, (with or without nuts), bolts (with nuts), rivets, and pins.

FILE: A patterned cutting tool in the shape of a bar, usually held by hand, which is employed to remove stock from metal parts.

FILLET WELD: A weld placed at the inside intersection of two perpendicular parts.

FINE BLANKING: A press work process that uses tremendous pressure to stamp flat parts from sheet. Unlike punching, the piece cut from the plate is the desired part, and the remaining plate is waste.

FLY CUTTER: An adjustable tool used to cut circles in metal parts. When used to cut ring-shaped grooves, the process is called trepanning.

FORGING: Heating a metal part to a high temperature and forming it with hammers or hydraulic dies.

FORKLIFT: A vehicle equipped with a pair of forks which can be raised or lowered hydraulically to lift heavy parts.

FULCRUM: The third component in controlled bending, the first two being a force and a counter force. The shape of the fulcrum is assumed by the bent object.

GALVANIC CORROSION: A flow of current caused when two dissimilar metals are in contact via an electrolytic solution, which results in the corrosion of one of the metals. See the galvanic corrosion chart on page 20.

GALVANIZING: A hot dip process of plating in which a part is dipped into a molten vat of zinc, which adheres in a thin layer to its surface. As the zinc cools, it forms a crystalline structure called "spangles."

GANTRY: A movable frame structure from which a hoist or other device can be mounted.

GEARS: A wheel with grooves or teeth cut in its surface. The grooves/teeth are designed to mesh with teeth in other gears and transmit motion.

GRINDER (ELECTRIC OR AIR): A handheld tool used to hold abrasive products (like discs), stones, or burs.

HACKSAW: A handheld tool that holds a tensioned, serrated blade typically used to cut extrusions.

HOIST: A device that uses gears or pulleys to provide mechanical advantage in lifting. Some hoists are equipped with brakes to prevent slippage and with electric motors to provide power.

HONING: The process of grinding rough holes to make them dimensionally precise and smooth.

HOT DIPPING: A process for coating metal parts with a layer of another molten metal. The base part is dipped into a vat of molten metal, which adheres to its surface. Hot dipping with zinc is called galvanizing; dipping in aluminum is called aluminizing.

HOT-ROLLED STEEL: Steel produced by rolling or extruding while in a molten state. Hot-rolled steel is less dimensionally accurate than cold-rolled steel, has rounded corners, and a thin coating of mill scale with a blue, purple, or brown surface coloration.

HEADSTOCK: The end of a lathe which is turned by the motor.

HYDRAULIC RAM: A hardened shaft powered hydraulically that is used to force dies into metal parts to change their shape.

I BEAM: A generic name for structural extrusions shaped in cross section like a serif capital "I" and composed of two parallel flanges and a perpendicular web at center. I beams are categorized by their intended use into W, S, M, and HP shape categories.

INCRALAĈ: A clear acrylic lacquer designed for coating copper and copper alloys in exterior applications.

INERT GAS: A non-reactive gas used in welding to shield the weld zone from the atmosphere; argon is commonly used for this purpose.

INGOT: A rough bar of metal.

ION: An atom or molecule that has gained a net electric charge by acquiring or losing electrons.

JACK: A hydraulic or ratcheting device used to lift heavy objects.

JIG: A device used to clamp parts in particular positions during assembly. Jigs are usually made for multiple runs of parts, to speed assembly and maintain dimensional consistency.

JIGSAW:	A handheld cutting tool that uses an oscillating, serrated blade to cut the workpiece.
KERF:	A cut; specifically, the thickness of a cut.
KNURLING:	The application of a grooved pattern to metal parts, primarily to make them easier to grip.
LACQUER:	A clear organic coating consisting of a solution of shellac, resin, cellulose, or polymers mixed with an evaporating liquid solvent.
LAPPING:	The use of very fine abrasive pastes to create dimensionally accurate, flat surfaces.
LASER CUTTER:	A device that uses a laser beam to cut through a wide variety of materials.
LATHE:	A machine tool on which cutting tools are applied to a spinning workpiece.
LEVELERS:	Bumpers attached to a threaded stem which can be placed on the bottom of furniture and equipment to compensate for uneven floor surfaces.
LUBRICANTS:	A material, such as grease or oil, that is applied to moving parts to reduce friction. Lubricants are also used in most cutting, drilling, and machining operations to keep the cutting tool and the workpiece cool and prevent damage to either.
MACHINE SCREW:	Threaded fasteners with a wide variety of head configurations designed for relatively low-strength assembly applications.
MANDREL:	In the case of tube bending, a mandrel is a flexible device inserted into the tube to keep it from deforming in cross section during the bending process.
MASTER (SPINNING):	A wooden or metal form attached to the spinning lathe against which the blank is forced to conform.
MIG WELDING:	*See* electric arc welding.
MILL:	A machine tool that uses serrated cutters fed spinning into the workpiece to remove stock. The workpiece is attached to a table that can be precisely moved in the x, y, and z axes in relation to the tool, thereby making extreme accuracy possible.
MILL SCALE:	Flaky, blistered layers of oxide formed on the surface of steel during hot working processes (e.g., forging, hot rolling). Unlike rust, which is bright orange or red, mill scale is usually bluish, brown, or purple.
MITER CUT:	An angular cut made in an extrusion to join two parts oriented in different directions.

NECKING:	A press work process used to shrink the diameter of a section of a tubular part.
NEOPRENE:	A synthetic rubber material.
NESTING:	Locating multiple parts to be cut from a sheet or plate in the configuration that will yield the most efficient use of the material. Originally a trial and error process performed graphically, nesting calculations are now typically performed with the aid of a computer.
NON-WOVEN ABRASIVES:	A nylon fiber pad coated with abrasive particles suspended in resin.
NOSE BAR:	The part on a brake that serves as a fulcrum.
NUT:	An internally threaded fastener that, when combined with a screw or a bolt, creates a tight clamp that holds two or more parts together.
OIL CANNING:	Warping and buckling that occurs on thin sheet panels when they are unable to resist shear.
ORE:	Naturally occurring mineral deposits that contain significant quantities of a desired metal. The ore is crushed and a variety of processes are used to separate the metal from unwanted impurities. *See* smelting.
ORGANIC COATINGS:	Continuous coatings such as paints, varnishes, enamels, and lacquers.
OXIDATION:	Formation of oxides on the surface of a metal part; in steels, this appears in the form of rust or mill scale.
OXIDE:	A compound formed by the combination of oxygen and an element.
OXYGEN FLAME CUTTING:	The use of a controlled flame fed by a mixture of oxygen and a fuel gas to create a melt zone. By moving the torch, a linear cut is produced.
OXYGEN/FUEL WELDING:	Similar to oxygen flame cutting, but the flame is used to melt the edges of two parts to be welded together. In the melt zone (also called a puddle), a certain amount of molten metal, called filler, is added. The mixture solidifies as it cools, creating a strong joint.
PALLET:	A platform made from wood studs or plywood and used as a support for large or heavy parts in transit.
PALLET JACK:	A hand-operated rolling cart with a pair of forks that fit into pallets and lift them slightly to allow their transport along a shop floor.
PAWL:	A small curved part that prevents a special gear, called a ratchet, from reversing its motion.
PERFORATED SHEET (PERF):	Metal sheet that has been punched with a pattern of holes (usually round). If the holes are arranged diagonally (at 60 degrees to each other), the pattern is called staggered; if they are spaced in a 90-degree grid, the pattern is called straight.

PHOSPHATE COATING: A dipping process used on parts prior to application of primer or organic coating to improve corrosion resistance and create a microscopically "rough" surface, good for bonding to primer.

PICKLING: A chemical process performed (usually on hot-rolled steel) to remove mill scale and rust.

PILLOW BLOCK: A housing that holds a radial (ring-shaped) bearing. The housing has holes for mounting the pillow block securely. Pillow blocks are typically meant for use in pairs, with a rotating shaft suspended between them.

PIPE: A hollow extrusion that is nominally graded by its inside dimensions (i.d.). Pipe is primarily intended for structural applications and the carrying of flowing liquids and gases, sometimes under pressure.

PISTON: A shaft designed to transmit motion along its main axis, which is held in a cylinder and is pushed and released by hydraulic force or compressed gas.

PLASMA ARC CUTTING: A process that employs a stream of compressed air, heated to plasma, and blown through the cut zone.

PLATE: A thick sheet of metal.

PLATING: One of a variety of processes used to coat parts with a thin layer of metal.

PLUG WELD: A weld produced by laying one plate over another and filling a hole in the upper plate with material; also called a rosette weld.

POLISHING: The process of sequentially using finer and finer abrasives to produce a smooth surface on a metal part with more or less specular (reflective) properties.

POWDER COATING: An electrostatic process in which electrically charged dry paint particles are fired at a metal part, which is grounded and attracts an even layer of the particles. The coating is then baked to form a continuous layer.

PRIMER: The first layer of protection placed on a part prior to undercoating and the application of an organic top coating.

PUNCH: A tool that uses extremely focused force to push hardened dies through sheet and plate to produce holes. The material knocked out of the hole is waste; the reverse process is called fineblanking.

QUENCHING: Heating and then cooling of steel parts in a controlled bath (usually water, brine, or oil) to alter the hardness of the part.

RATCHET: A special gear designed to turn in one direction only. *See also* pawl.

REAMER: A cutting tool used to enlarge drilled holes.

RESISTANCE WELDING (SPOT WELDING):	A process used to weld thin sheet together by clamping the sheets between electrodes. The resistance caused when current is run between the electrodes creates a small localized melt, which bonds the sheets.
RIVET:	A fastener used primarily for conditions where only one side of the assembly is accessible.
ROSETTE WELD:	*See* plug weld.
RUST:	A bright-red or orange layer of oxide that can form on the surface of iron or steel. Rust can eventually destroy steel parts if the process of oxidation is not halted. *See* oxidation.
SANDBLASTING:	A process using silica sand propelled by compressed air to blast rust, mill scale, and paint from the surface of metal parts.
SCREW:	A threaded fastener designed for use with a nut or insertion into a tapped hole. The head of a screw is designed to be turned by a wide variety of driving tools.
SET SCREW:	A threaded, headless fastener typically used to fasten parts using compression.
SHAFT:	A spinning rod designed for the transmission of motion.
SHEAR:	A tool that uses a pair of hardened blades to make cuts in sheet and plate.
SHEET:	Flat, relatively thin panels of metal graded by thickness.
SHOT:	Tiny spheres of metal (usually lead or steel), graded by size and used for a variety of purposes, including blasting and peening.
SHOT PEENING:	A process using steel shot to condition the surface of metal parts. Shot peening is used both for its decorative quality and its ability to impart extra rigidity to the surface of parts by creating millions of tiny dimples.
SLIP ROLLER:	A tool used to give metal products a continuous curvature by forcing them between a series of polished rollers.
SMELTING:	The reduction of ore to pure metal. The ore is crushed, mixed with a fuel source such as coke, stoked with air, and burned. When the melting point of the metal is reached, it coalesces into a liquid ball at the bottom of the mix.
SNIPS:	Handheld tools used like scissors to cut sheet metal and remove burrs.
SOLVENT:	A solvent is a liquid that dissolves another substance.
SPINNING:	The process of forcing flat sheet metal discs against spinning masters on a lathe to create parts with compound curvature.
SPRAY METALLIZING:	A method of plating parts with metal. Tiny particles of molten metal are sprayed through a gun at an armature, which may or may not be made of metal. The particles fuse and cool, forming a continuous layer.

SPRING:	An elastic device that returns to its original shape when not under load.
SPUR GEAR:	A thick, star-shaped gear that rolls like a wheel along a linear, toothed track, called a spur rack.
STEEL:	The primary metal used for fabrication. Steels are iron alloys produced in a staggering variety for different applications. Carbon steels are the most common category. Alloy steels encompass most specialty alloys, including stainless steels.
STICK WELDING (SMAW):	*See* electric arc welding.
STOKING:	Feeding oxygen into a blast furnace to produce hotter combustion.
STONES:	Blocks of an abrasive/resin mixture formed into a variety of shapes and mounted to shafts for use with grinding tools.
STUD WELDER:	A tool used to weld threaded studs to the surface of metal parts.
TAILSTOCK:	The stationary end of a lathe.
TAP, TAP WRENCH:	A tap is a hardened steel tool with tapering, spiraling cutters. It is inserted into a drilled hole to cut threads for screws. Taps can be used manually with a tap wrench, which is a small, handheld vise.
TEE:	A T-shaped extrusion consisting in cross section of a horizontal flange and a vertical stem. Structural steel tees are frequently made in steel by splitting I beams at their mid section.
TEFLON®:	A synthetic material used in applications where its low coefficient of friction is valuable.
TEMPERING:	Heating metal and then cooling it in a controlled manner to change its hardness.
THREAD:	Matching internal or external spiral grooves cut into a fastener or a part to create mechanical fastening.
TIG WELDING:	*See* electric arc welding.
TREPANNING:	*See* fly cutter.
TUBE:	Hollow metal extrusions graded by their outside dimensions (o.d.). Tube is available in a wide variety of profiles, the most common of which are round, square, and rectangular.
TURRET:	The top of a mill containing the motor, gearing, and the spindle.
UNIVERSAL JOINT:	A linkage between shafts which allows for the transmission of rotary motion off axis (not in parallel alignment) between shafts.
UPSETTING:	A press work process used to create a "mushroomed" head on solid stock, such as the head of a screw.

VEE GROOVE: A beveled "valley" created between two parts to be welded in order to provide a space for molten material to be deposited.

VIBRATORY DEBURRING: A process used to remove burrs and condition the surfaces of small parts. The parts are placed in a tank with tumbling media (pellets made from stone or ceramics), and the mixture is vibrated, causing the media to gently strike the metal parts repeatedly.

VISE: A stationary tool used to securely hold a workpiece while operations are performed on it.

VITREOUS ENAMEL: A non-organic coating that consists of a paste of crushed glass, which is spread on the part to be coated, melted at a high temperature, and allowed to cool and bond to the surface of the part.

VOLATILE ORGANIC COMPOUNDS (VOCS): Undesirable chemicals released into the atmosphere by solvents in paint and other organic coatings. VOCs are the subject of substantial governmental regulation in the United States.

WASHER: A component in a screw or bolt assembly; a circular ring placed between a nut and the head of a screw or bolt and the fastened substrate.

WATER JET CUTTING: A cutting process that uses a highly concentrated steam of water mixed with abrasive particles to cut metal and other materials.

WIRE MESH (ALSO WIRE CLOTH): Stock component made by weaving various thicknesses of metal wire to form a metallic "textile."

WORM GEAR: A threaded gear used to transmit motion directly from a spinning shaft.

WROUGHT IRON: An iron alloy with a very low carbon content that was used as the primary metal for making objects by blacksmiths prior to the ascendancy of carbon steel. Praised for its "workability" in comparison with modern carbon steel.

X, Y, AND Z AXES: The three axes used to describe three-dimensional movement on machinery; on a vertical mill, the x and y axes are perpendicular to each other and parallel to the mill table. The z axis is perpendicular to the table and parallel to the spindle (up and down).

ZEE: A Z-shaped extrusion consisting in cross section of two horizontal flanges, pointing in opposite directions and attached to a vertical web. Zees are fairly uncommon and stocked in limited sizes, when they can be found.

INDEX

A

Abrasives, 84, 101–5, 111
Accuracy in design
process
and computer
numerical control
(CNC), 158, 161
specification guidelines
for, 139–43
Acme threads, 150, 151
Alloys, 11, 13–15, 17–18
Aluminum, 16–17, 22,
23, 25–26, 73, 108,
124, 129
American National
Standard for United
Screw Threads, 89
Annealing, 15
Anode, 125, 128
Anodizing, 111, 129–30
Architectural extrusions,
26
Armatures, 131–36
Art Deco metalwork, 77

B

Ball bearings, 145
Ball burnishing, 110, 115
Band saws, 41
Bar stock, 26–28, 39
Bauxite, 16
Bearings, 145
Belts, 104
Bench grinders, 102
Bending, 50–51, 161
Binders, 117
Bits, 75–76
Blacksmithing, 9, 15
Blanks, 58
Blending, 98–105
Blind fastening, 96–97
Bolts, 87–89, 92
Brakes, 54–55
Brass, 17–19, 23, 28,
119, 120

Brazing, 86
Broach, 42, 43
Bronze, 17–19, 23, 28,
119, 120
Brushing, 111–13
Bulging, 63
Burs, 31, 84, 99, 101
Bus bar, 28

C

Camber, 25
Cap screws, 93
Carbon steel, 13, 15–16
Case hardening, 15
Casters, 146, 147
Casting, 9
Channels, 25
Chemical milling, 73
Chop saws, 39
Chromium, 14
Chucks, 45–46, 47, 64,
66, 75
Circle cutting, 38–39,
45, 63
Clearcoating, 119–20
Coatings, organic,
117–23
Coining, 59
Cold hardening, 15
Cold rolling, 21–22, 24,
28
Cold saws, 40
Compression springs,
151
Computer aided drafting
(CAD) programs,
159–60, 161
Computer numerical
control (CNC),
158–61
bending, 161
continuous path
control, 160
cutting, 33, 34
lathes, 77

mills, 70
nesting, 160–61
Continuous path
control, 160
Copper, 11, 17–19, 23,
28, 119, 120, 126
Corrosion, 13–14,
19–20, 117, 128
COR-TEN steel, 14
Cotter pins, 97
Counter bores, 44–45
Countersinks, 44, 45
Cranes, 154
Curling, 59
Cutting and shearing,
31–41, 160–61

D

Deburring, 104, 108–10,
115
Deep drawing, 58–59,
61, 62
Diamond plate, 23
Die grinders, 99, 102
Dies, 90, 160
Disc sanders, 104
Dollies, 154
Dressers, 102
Drill bits, 42, 44, 45
Drilling, 42–48
Drill press, 46–48

E

Elastic phase, 50
Electric arc welding,
79–82, 86
Electrolytes, 125, 129
Electroplating, 125–28
Electropolishing, 115
Embossing, 58, 59, 73
Enamels, 117
End mills, 67, 72
Etching, 73
Expanded metal, 24
Extension springs, 151

Extrusions, 24–25
bending, 51–53
cutting, 39–41
product types, 25–30
rolling, 55–56
specification
guidelines, 138

F

Fabrication centers, 49
Fasteners, mechanical,
87–97, 133, 135, 152
Feed screws, 70
Files, 98–99
Fillet welds, 84, 85
Fine blanking, 59
Flatbed trucks, 157
Fluxes, 78
Fly cutters, 44, 45, 73
Forging, 59
Forklifts, 153–54
Frame-supported shapes,
131–36
Friction welding, 83
Fulcrum, 50

G

Galvanic corrosion,
19–20, 136
Galvanizing, 119,
128–29
Gantry, 72, 154
G code programs, 159
Gears, 147–50
Grinding, 98–105
Grinding stones, 102
Groove welds, 84, 85

H

Hacksaws, 31, 32, 39
Hand brakes, 54–55
Hand drills, 43, 46
Headstock, 75, 76
Hinges, 151, 152
Hoists, 154

Hole saws, 44, 45
Hot dipping, 128–29
Hot rolling, 21, 24, 28, 30
HSLA steel, 14
Hydraulic ram, 59, 60, 61
Hydroabrasive (water-jet) cutting, 34, 35

I
I beams, 25
INCRALAC, 119–20
Iron, 11–12

J
Jacks, 150, 153
Jigsaws, 31, 32, 33

K
Kerf, 31, 33, 34
Knurling, 76, 142

L
Lacquers, 117, 120
Laser cutting, 34, 36, 37, 38
Laser polishing, 115
Lathe dog, 75
Lathes, 73–77
Lead, 19, 118
Lift gates, 157
Lock washers, 95, 152
Lubricants, 144

M
Machine screws, 92–93
Machining, 67–77
Mandrels, 53, 55, 97
Metal products, 21–30, 138
See also Extrusions; Sheets
MIG welding, 80–81
Mills, 67–73
Moldings, 141
Motion-enabling devices, 144–52

N
National Association of Architectural Metal Manufacturers, 115
Necking, 63
Nesting programs, 160–61

Notchers, 39
Nuts, 87–88, 93–94, 133, 150

O
Oil canning, 131
Oxidation, 20, 78, 106, 117, 128, 129
Oxygen-flame cutting, 33
Oxygen/fuel welding, 78–79

P
Painting, 117–23
Pallet jacks, 153
Pallets
loading, 157
moving, 153–55
Passivation, 118–19
Patinas, 19, 119–20, 126
Patterned tools, 98–101
Pawl, 149
Perforated (perf) sheet, 23, 24
Photoresist, 73
Pillow blocks, 145, 146
Pins, 97
Pipe, 28–30, 89
Pistons, 150, 151, 152
Plasma-arc cutting, 33–34, 38, 49
Plastic phase, 50
Plating, 124–29
Plug (rosette) welds, 84, 85
Polishing, 102, 113–16, 142
Porcelain (vitreous) enamel, 123
Powder coating, 121–23
Press brakes, 55, 58, 59, 60
Press work, 58–63, 160
Primer coats, 119
Pulleys, 148, 149
Punching, 37–38, 48–49

Q
Quenching, 15

R
Rail profiles, 25
Random-orbital sanders, 102–4
Ratchets, 149–50

Reamers, 42, 43, 45
Resins, 117–18
Resistance welding, 82–83
Ring shears, 38–39
Rivets, 96–97
Roller tracks, 157
Rolling, 50–51, 55–57
Rosette (plug) welds, 84, 85, 133
Rotary punch, 49
Rust, 20, 117

S
Sandblasting, 106–8
Sanders, 102, 103, 104, 111–13
Screws, 87–89, 92–93, 133
Security screws, 93
Self-supporting shapes, 136–37
Set screws, 92
Shape cutting, 31–38
Shears, 37–38, 39
Sheets
bending, 54–55
cutting, 31–38
press work, 58–63
product types, 21–24
rolling, 55
specification guidelines, 138
Shock absorber, 152
Shot blasting (peening), 108
Slip rollers, 55
Smelting, 11–12
Snips, 31
Soldering, 86
Solvents, 118
Spangles, 129
Specification guidelines, 138–43
Spinning, 61, 63–66, 136–37
Spot welding, 82–83
Sprayers, paint, 121
Spray metallizing, 129
Springs, 151–52
Sprockets, 147
Spur gears, 147, 148
Stainless steel, 13–14, 21–22, 28, 73, 106, 115, 118–19, 120
Stamping, 58

Steel
properties of, 12–16
sheet, 22, 26, 138
See also Stainless steel
Stick welding, 81–82
Stoking, 11–12
Structural extrusions, 25–26
Stud welding, 95
Swing, 75

T
Tailstock, 75
Taps, 44, 45, 90–91
Tap wrenches, 90–91
Tee sections, 25
Threads, 87, 88, 89–91
TIG welding, 81, 83
Timesaver, 111–13
Titanium, 19
Tool steel, 15
Torsion springs, 151
Trepanning, 73
Tubing, 29–30, 52–53
Turret, 67, 69, 72

U
Ultrasonic welding, 83
Universal joints, 147
Upsetting, 59

V
Vacuum lifters, 157
Varnishes, 117
Vitreous (porcelain) enamel, 123
Volatile organic compounds (VOCs), 118
Volumetric solids, 131–37

W
Washers, 89, 94–95, 136, 152
Water-jet cutting, 34, 35, 38
Welding, 78–86, 133, 135, 136
Wheels, 145–46
Worm gears, 149
Wrought iron, 12, 15

Z
Zee, 25
Zinc, 118, 119, 126, 128–29